Amitavo Roy Choudhury

Investigations of after cavity interaction in gyrotrons including the effect of non-uniform magnetic field

Karlsruher Forschungsberichte aus dem
Institut für Hochleistungsimpuls- und Mikrowellentechnik

Herausgeber: Prof. Dr.-Ing. John Jelonnek

Band 4

Investigations of after cavity interaction in gyrotrons including the effect of non-uniform magnetic field

von
Amitavo Roy Choudhury

Dissertation, Karlsruher Institut für Technologie (KIT)
Fakultät für Elektrotechnik und Informationstechnik, 2013
Referenten: Prof. Dr. rer. nat. Dr. h.c. Manfred Thumm,
Prof. Dr.-Ing. Dr. h.c. Klaus Schünemann

Impressum

Karlsruher Institut für Technologie (KIT)
KIT Scientific Publishing
Straße am Forum 2
D-76131 Karlsruhe

KIT Scientific Publishing is a registered trademark of Karlsruhe Institute of Technology. Reprint using the book cover is not allowed.

www.ksp.kit.edu

Print on Demand 2014

ISSN 2192-2764
ISBN 978-3-7315-0129-9

Foreword of the Editor

The rapid growth of world population and economy requires new methods of environmentally friendly energy generation. One future candidate is controlled thermonuclear fusion in magnetically confined plasmas where high-power microwave sources (gyrotrons) are used for plasma heating and stabilization to achieve the required temperatures of around 120 Million degree Celsius. Gyrotrons are electron cyclotron masers driven by weakly relativistic helical electron beams, operating in a longitudinal magnetic field and highly oversized cavities. The current state-of-the-art for frequencies up to 170 GHz is 1 MW output power in continuous wave (CW) operation at 50 % efficiency, including single-stage depressed collector operation. Gyrotrons have to run in single-mode operation in order to guarantee high output mode purity for low-loss transmission to the plasma torus and to generate sufficient low level of internal stray radiation. This mandatory requirement forces gyrotron developers to have a close look on to all possible parasitic oscillations in the tube. Experiments at KIT have shown that under certain conditions, parasitic oscillations are excited in the region after the main interaction region of the gyrotron. Those oscillations are named "After Cavity Interactions (ACI)". In most publications, ACI is described as stationary, hence, defined as "static ACI", That static ACI is characterized by the reduction of the overall gyrotron efficiency caused by an unwanted energy exchange after the cavity between the electron beam and the RF field.

In the present work Dr.-Ing. Amitavo Roy Choudhury considered the effect of "dynamic ACI". It is defined as the time-variant interaction of the electron beam either with the main mode or with different other competing modes, mainly at lower frequencies and, again, at a physical location behind the cavity. As prerequisite of his work, Dr. Choudhury modified the self-consistent multi-mode code SELFT. Doing that, he did extend the simulation domain of the KIT code into the up-taper region of the gyrotron. At this region the strength of the longitudinal magnetic field is strongly varying and has to be added to the simulation. The modified code has been applied to dynamic ACI studies using four different gyrotron configurations (two conventional and two co-axial cavities). The results of the simulation confirm the existence of ACI. A comparison with measurements results verifies the theoretical predictions.

Dr.-Ing. Amitavo Roy Choudhury provides the worldwide first detailed investigation of dynamic ACI in high-power megawatt-class gyrotrons. The results confirm that both up-taper radius contour and magnetic field profile have to be optimized in order to minimize ACI.

Investigations of After Cavity Interaction in Gyrotrons Including the Effect of Non-uniform Magnetic Field

Zur Erlangung des akademischen Grades eines

DOKTOR-INGENIEURS

von der Fakultät für
Elektrotechnik und Informationstechnik
des Karlsruher Institut für Technologie (KIT)

genehmigte

DISSERTATION

von

M.Tech. Amitavo Roy Choudhury

geb. in Balurghat, West Bengal, India

Tag der mündlichen Prüfung: 26.06.2013

Hauptreferent: Prof. Dr. rer. nat. Dr. h.c. Manfred Thumm
Korreferent: Prof. Dr.-Ing. Dr. h.c. Klaus Schünemann

Dedicated

To

My Beloved Parents

Smt. Anjana Roy Choudhury

&

Sri. Subhendu Roy Choudhury

Kurzfassung

Die steigende Nachfrage nach leistungsstarken, verlässlichen und effizienten Gyrotron-Oszillator-Röhren als Millimeterwellenquellen für Elektron-Zyklotron-Resonanz-Heizung (ECRH) und nichtinduktiven Stromtrieb in Fusionsplasmaexperimenten erfordert eine genaue Betrachtung aller möglicher parasitärer Schwingungen im Gyrotron. Dabei wurden neue Effekte beobachtet, wobei experimentell herausgefunden wurde, dass einige schwache Oszillationen auch im Bereich nach dem Gyrotron-Wechselwirkungsresonator angeregt werden. Deshalb werden diese Effekte als After-Cavity-Interaction (ACI) bezeichnet. In den meisten Veröffentlichungen wird die ACI als eine stationäre Reduktion des Gesamtwirkungsgrads beschrieben, die aufgrund eines weiteren Energieaustauschs zwischen Elektronenstrahl und Hochfrequenzfeld zustande kommt und hier als statische ACI bezeichnet wird. In der vorliegenden Arbeit betrachten wir die dynamische ACI, welche sich unterschiedlich verhält: die Wechselwirkung mit der Hauptmode oder mit anderen Moden bei niedrigerer Frequenz nach dem Resonator führt zu einer dynamischen Modulation der Ausgangsleistung.
Der selbstkonsistente KIT Multimoden-Code SELFT wurde modifiziert, um den Simulationsbereich in die Radiusaufweitungs-Sektion (Uptaper) des Gyrotronresonators zu erweitern, wo die Wahrscheinlichkeit des Auftretens dynamischer ACI besteht und die Stärke des Längsmagnetfeldes nicht mehr konstant ist. Dieser modifizierte KIT SELFT Code wurde zur Untersuchung der ACI in vier verschiedenen Gyrotronkonfigurationen eingesetzt (zwei Gyrotronröhren mit konventionellem zylindrischem Resonator und zwei Koaxial-Gyrotrons). Die Studie bestätigt, dass unerwünschte Wechselwirkungen in der Uptaper Region zu zusätzlichen parasitären Schwingungen mit Leistungen im 1%-Bereich führen können. Aufgrund der Einführung eines $dBR(z)/dz \neq 0$-Terms und der signifikanten Verkleinerung des Magnetfelds in Richtung Uptaper-Ausgang konnten einige schwache Störschwingungen im Hochfrequenzspektrum des Ausgangssignals beobachtet werden. Theorie und Messungen stimmen gut überein was die im SELFT Code durchgeführten Modifikationen bestätigt. Die Ergebnisse dieser Studie bestätigen, dass die Radiuskontur des Uptapers und das Profil des Magnetfelds optimiert werden müssen um die ACI zu minimieren.

Abstract

The increasing demand for powerful, reliable and efficient gyrotron oscillators as millimeter-wave sources for Electron Cyclotron Resonance Heating (ECRH) and non-inductive current drive in fusion plasma experiments calls for a close look on all possible parasitic oscillations in gyrotrons. New effects are observed where it has been experimentally found that some spurious oscillations are also excited by the electron beam in the region after the gyrotron interaction cavity. Therefore these effects are called After Cavity Interaction (ACI). In most publications, ACI is described as a stationary reduction in the overall efficiency caused by further energy exchange after the cavity between electron beam and RF field which here is named static ACI. In this study we considered dynamic ACI, which behaves differently: the interaction after the cavity, either with the main mode or with different modes at lower frequency, leads to a dynamic modulation of the output power.

The self-consistent KIT multi-mode code SELFT has been modified in order to extend the simulation domain into the radius uptaper section of the gyrotron cavity where the probability of occurrence of dynamic ACI exists and the strength of the longitudinal magnetic field is no longer constant. This modified KIT SELFT code has been applied to dynamic ACI studies using four different gyrotron configurations (two conventional cylindrical cavity gyrotrons and two coaxial cavity gyrotrons). The study confirms that undesired interactions in the uptaper region can result in additional parasitic oscillations with power in the 1% region. Due to the implementation of a $dB_R(z)/dz \neq 0$ term and the significant reduction of the magnetic field towards the uptaper output, some spurious oscillations have been observed in the RF spectrum of the output signal. Theory and measurements are in good agreement verifying the modifications performed on the SELFT code. The results of this study confirm that uptaper radius contour and magnetic field profile have to be optimized in order to minimize ACI.

Contents

List of used Symbols and Variables

Frequently used indices

$\perp$	perpendicular or transverse direction
$\parallel$	parallel or longitudinal direction
i	coaxial insert
j	counting index of ensemble of electrons
$i.e.$	that is
L	at the launcher
k	mode index
m,p	azimuthal and radial indices
n	axial mode index
R	at the cavity (resonator)
S	on the surface S
$*$	conjugate conversion of a complex function
Δ	Laplace operator
o	'outer'
dif	based on the radiated power from the resonator
$elec$	based on the output power from the electron beam
in	the cavity entrance (the side facing the gun)
$loss$	based on total losses
max, min	maximum or minimum value
out	at the resonator output
$\Delta_\perp$	transverse Laplace operator

List of Variables

$\vec{a}$	particle acceleration term
A	cross-sectional area of the resonator
$\vec{A}_{m,k}$	amplitude factor of Bessel functions of the second kind
b	magnetic compression with $b = B_R/B_E$
B_R	flux density of the static magnetic field at resonator
B_E	magnetic field at the emitter

B_0	maximum magnetic field at the resonator
$\mathrm{B}_{R,r}, \mathrm{B}_{R,z}, \mathrm{B}_{R,\varphi}$	flux density of the static magnetic field along radial, axial and azimuthal direction
$\vec{\mathrm{B}}$	vector of the magnetic flux density
$\vec{B}_{m,k}$	amplitude factor of Bessel functions of first kind
c	velocity of the light (2.998×10^8m/s)
$C_k, C_{m,p}$	normalizing factor
$\vec{\mathrm{D}}$	vector of electrical displacement density
e	charge of the particle (1.6×10^{-19}C)
$\vec{\mathrm{e}}_k$	electric field vector of the transverse field components of a mode 'k'(eigenvector)
$\vec{\mathrm{e}}_k^+$	eigenwaves in backward direction
$\vec{\mathrm{e}}_k^-$	eigenwaves in forward direction
e_z, e_r, e_φ	unit vectors of the different coordinate systems
$\vec{\mathrm{E}}$	vector of the electric field strength
$\mathrm{E}_\perp$	electric field in transverse direction
E_r, E_φ, E_z	electric field in radial,azimuthal and axial direction
f	frequency
C_F	fresnel parameter
F	normalized field amplitude
f_c	relativistic electron-cyclotron frequency $f_c = \Omega_c/2\pi$
$\vec{f}_k(z,t)$	transverse component of complex field profile of a mode 'k'
$\vec{f}_{\parallel,k}(z,t)$	longitudinal component of complex field profile of a mode 'k'
$\vec{f}_k^+$	complex field profile of the backward wave of mode 'k'
$\vec{f}_k^-$	complex field profile of the forward wave of mode 'k'
$G_{e,k}$	coupling factor
$\vec{\mathrm{H}}$	complex magnetic field strength vector with $\vec{\mathrm{H}} = \vec{\mathrm{B}}/\mu_0$
$\mathbf{I}$	identity matrix
I_B	beam current
$I_{B,j}$	beam current of each macro-particle
I_{lim}	limiting current for $\beta_{\parallel} \rightarrow 0$
j	imaginary unit $\sqrt{-1}$
J_E	emitter current density
$\vec{J}$	complex current density vector
J_m	Bassel function
J_{sc}	space charge current density

k	wave number $k = 2\pi/\lambda = 2\pi \cdot f/c$
$k_\perp$, $k_{\|\|}$	perpendicular and axial wave number $k^2 = k_{\|\|}^2 + k_\perp^2$
L_d and L_u	length of down and uptaper section of the cavity
L_R	length of cylindrical cavity section
$L_{s,d}$ and $L_{s,u}$	length of parabolic smoothing towards up and downtaper
m	electron rest mass ($m = 9.11 \times 10^-31$kg)
N_m	m-th order Neumann function
p	total output power from the cavity
$\vec{p}$	slowly varying complex electron momentum vector
$p_\perp$	transverse electron momentum
p_{dif}, p_{ohm}	power due to radiation and ohmic loss
Q_{dif}, Q_{ohm}	diffractive quality factor and ohmic quality factor
Q	total quality factor with $1/Q = 1/Q_{dif} + 1/Q_{ohm}$
Δq	charge
r_L	*Larmorradius*
$\vec{R}_k$	excitation term of a mode 'k'
R_i	radius of coaxial insert
R_0	outer wall radius of the resonator
R_e	guiding center radius of an electron orbit
s	harmonic number
t	time variable
t_s	simulation time in ns
Δt	time step
u	normalized particle velocity
$\mathrm{u}_\perp$	normalized transverse velocity component
$u_{\|\|}, u_r, u_\varphi$	normalized longitudinal, radial and azimuthal velocity component
U_B	beam voltage
U_k	cathode voltage
ΔU	voltage depression ($\|\Delta U\| = \|U_k - U_B\|$)
v_{ph}	wave phase velocity
v	particle velocity
$v_\perp, v_{\|\|}$	perpendicular and parallel velocity components
v_b	beam voltage in term of eV ($U_B - \Delta U$)
W	stored energy
W_{kin}	kinetic energy

z	axial z-coordinate
Δz	discretization step size in the z direction
α	velocity ratio (pitch factor) with $\alpha = v_\perp / v_{\shortparallel} = \beta_\perp / \beta_{\shortparallel}$
β	normalized velocity with $\beta = v/c$ and $\beta^2 = \beta_\perp^2 + \beta_{\shortparallel}^2$
β_{in}	initial normalized velocity β
$\beta_\perp, \beta_{\shortparallel}$	electron normalized velocity perpendicular and parallel to the magnetic field
$\beta_{\shortparallel 0}$	$\beta_{\shortparallel}$ without considering the voltage drop
γ	relativistic factor (Lorentz factor)
γ_{in}	initial relativistic factor γ
Γ	complex reflection coefficient
δ	skin depth at frequency f for the cavity radius R_0
$\triangle$	normalized detuning factor
ε	dielectric constant
ε_0	dielectric constant of the vacuum (8.85×10^{-12} F$/m$)
η	efficiency
η_{out}	total efficiency
η_{elec}	electronic efficiency with $\eta_{elec} = \alpha^2 / 1 + \alpha^2 \cdot \eta_\perp$
$\eta_\perp$	transverse efficiency
λ	wavelength
$\lambda_{\shortparallel}$	wavelength in the waveguide
λ_0	free-space wavelength in vacuum
Λ	slowly varying phase of electrons in the rotating frame
μ	normalized interaction length or permeability
μ_0	permeability of free space with $\mu_0 = 4\pi \cdot 10^{-4}H/m$
ζ	phase shift in ψ or normalized axial position
π	$\approx$3.1416
$\vec{p}_{ohm}$	average ohmic power density
σ	material conductivity
τ	transit time of electrons through the resonator
φ	azimuthal angle in cylindrical coordinates
φ_e	azimuthal angle of the guiding center of the electron motion
Σ	summation
$\chi_{m,p}$	eigenvalue of a TE$_{m,p}$ mode
ψ	fast varying phase of the electron motion
ω	angular frequency$(= 2\pi f)$

ω_k	average frequency of the field of a mode 'k'
$\omega_{carrier}$	carrier frequency
ω_p	plasma frequency
$\omega_{\perp,k}$	cut-off frequency of a mode 'k'
Ω_0	non-relativistic cyclotron frequency with $\Omega_0 = eB/m$
Ω_c	relativistic cyclotron frequency with $\Omega_c = \Omega_0/\gamma$
Ω_f	average frequency of the particle
$\vec{\omega}$	complex resonance frequency of a mode 'k'in the stationary cold cavity field profile
ω_{RF}	frequency of RF oscillation
θ	varying gyro-phase of the electron
θ_{in}	electron input phase position

1 Introduction

Microwave tubes are still extensively employed in certain areas for applications for which their solid state semiconductor counterparts are not able to compete as far as delivering the required power level at microwave to millimeter wavelengths. These applications vary from Electron Cyclotron Resonance Heating (ECRH) of plasmas in fusion reactors to the sintering of high-performance industrial ceramics. This unique ability to produce hundreds of kilowatts of pulsed to continuous waves (CW) at frequencies higher than 200 MHz has made microwave tubes an indispensable source for high power and high frequency applications. There are various types of microwave devices which can be found in [Gil86]. Conventional microwave tubes impose limitations on the interaction circuit physical size and output power as the operating frequency increases. The interaction structure gets smaller with increasing frequency which consequently puts a limitation on output power [Bas96].
In order to meet the increasing demand for energy, it is necessary to have some alternative power plants, instead of having power plants using fossil fuels. Out of all the promising candidates, power plants using thermonuclear fusion in magnetically confined plasmas are the most promising one. Thermonuclear power plants have many advantages over the nuclear fission plants like there is no need for disposal of the spent nuclear rods, fuels can be recycled and also there is no CO_2 emission. In these thermonuclear plants the fusion process is started by heating a gas mixture, consisting of Deuterium and Tritium, up to several 100 Million degree of centigrade, in order to form a plasma. As a result a large number of neutrons are produced which transfer their kinetic thermal energy (80% of the total fusion energy) to a Lithium blanket so that a turbine generator can operate to produce electricity. An efficient way, by which the plasma heating is carried out, up to the required elevated temperature, is by using microwaves for electron heating which is accomplished by ECRH. This method has many advantages over other employed methods like it has a very high coupling factor into the plasma and also increases the possibility of localized heating. The minimum power level required for the generation of burning plasmas in future fusion reactors depends on the Q-factor. It is 50-100 MW and the required operating frequency range lies between 100 to 250 GHz. Such high frequencies are needed for efficient non-inductive current drive. Today, for this power and frequency regime gyrotrons are the only solution. A current overview of the state-of-the-art of gyrotrons, gyro-devices and free electron masers and their fields of application is given in [Thu13]. Thus, gyrotron operation with highest possible efficiency at the required frequency and power level and also with low spurious oscillations gain increasing importance for the successful realization of future reactors. The presented theoretical work is mainly concerned with the investigation of

spurious oscillations present in the output of gyrotron interaction cavities. It starts from the KIT developed time-dependent self-consistent slow-variables multi-mode interaction SELFT code.

1.1 Gyrotron components and principle of operation

The gyrotron is a fast wave microwave tube which emits radiation in the millimeter and sub-millimeter wavelength range. It covers the gap between conventional microwave tubes (for example klystrons, magnetrons, etc.) and far-infrared lasers [KBT04]. It is based on the Electron Cyclotron Maser (ECM) instability discovered in the late 1950s, when three researchers began to examine theoretically the generation of microwaves by the ECM interaction [FGP+77, HG77]: Richard Twiss in Australia [Twi58], the German Jürgen Schneider in the US [Sch59] and Andrei Gaponov in Russia [Gap59]. The same is based on the fact that moving electrons gyrate in the presence of an axial magnetic field due to the Lorentz force, hence the name gyro-tron. Latest gyrotron developments extend into the THz-regime [GLG08, BGI+09]. The schematic of a state-of-the-art high power gyrotron with a coaxial insert cavity and lateral RF output is shown in Fig. 1.1; its elements are explained below. A gyrotron consists of following major parts: Magnetron Injection Gun (MIG), magnet, beam tunnel, interaction cavity, quasi-optical mode converter, RF output window and collector.
Gyrotrons require a hollow electron beam with sufficient orbital velocity ($v_\perp$) and small velocity spread in order to generate high power RF radiation. The MIG produces an annular electron beam in which electrons execute small cyclotron orbits at a frequency required for cyclotron resonance interaction in the cavity. The magnet is used to provide the necessary magnetic field in the cavity while a compensation coil is used to suppress the field produced by the main coil in the emitter region. The MIG used in gyrotrons are usually operated in the temperature-limited (TL) region of the emitter [Pio93]. In addition to that, a sufficient parallel velocity component $v_{\shortparallel}$ is also required so that the electron beam can reach the interaction region and leave the cavity after the energy transfer. Due to the axial and the transversal component of the electron velocity in the presence of the magnetic field, the electron beam forms helical trajectories. It is noted that the radius of rotation of the electrons in their helical path i.e. *Larmorradius* $r_L = v_\perp/\Omega_c$ is much smaller than the beam radius . Ω_c is the electron cyclotron frequency. The ratio of the perpendicular velocity component $v_\perp$ to the parallel velocity component $v_{\shortparallel}$ is called *pitch factor*. It is defined as:

$$\alpha = \frac{v_\perp}{v_{\shortparallel}} \quad \text{with} \quad \beta_{\perp,\shortparallel} = \frac{v_{\perp,\shortparallel}}{c} \tag{1.1}$$

The velocities $v_\perp$ and $v_{\shortparallel}$ are generally normalized to the velocity of the light c. Thus total normalized electron velocity is given by:

$$\beta = \sqrt{\beta_\perp^2 + \beta_{\shortparallel}^2} \tag{1.2}$$

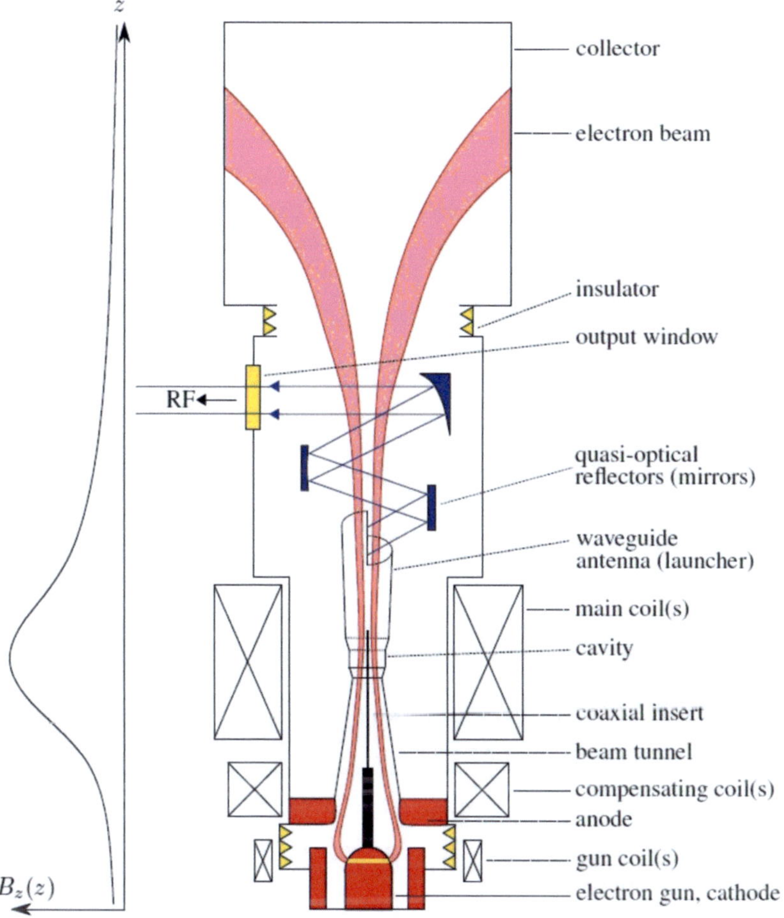

Figure 1.1: Schematic diagram of a gyrotron in coaxial cavity configuration.

The electronic efficiency of a gyrotron is given by:

$$\eta_{elec} = \frac{\alpha^2}{1+\alpha^2}\eta_{\perp} \tag{1.3}$$

where $\eta_{\perp}$ is the efficiency of the energy transfer from the perpendicular energy component of the gyrating electron beam to the generated RF output power. The maximum theoretical value is $\eta_{\perp} \approx 72\%$ [KDS+85], based on the theory of normalized variables (discussed in section 1.1.3). The value of $\eta_{\perp}$ depends primarily on the main operating parameters of the MIG i.e. beam current and beam voltage. The overall efficiency η_{tot} is even lower than the electronic efficiency $\eta_{\perp}$. Typically, it is about 30% if no other measures for energy recovery are used.
In the interaction cavity the energy exchange between electron beam and RF takes place due to the electron-cyclotron interaction mechanism. The purpose of the interaction cavity (either convention or coaxial type) is to provide a stable oscillation with one of the eigenmodes of the waveguide. The coaxial cavity consists of a simple cylinder with a down taper section of length L_d at the cavity entrance and an up taper section of length L_u at the cavity exit along with a longitudinally corrugated coaxial insert. For an in-depth description about the coaxial cavity see [Ker96]. The schematic diagram of a cavity with a tapered coaxial insert for a high power gyrotron along with a typical longitudinal field profile is shown in Fig. 1.2. In case of a conventional cavity, the coaxial insert is not used. The transitions between each section are normally smoothed, over a length of $L_{s,d}$ and $L_{s,u}$ for the downtaper and uptaper section respectively, using a parabolic function to minimize unwanted mode conversion at sharp edges. Gyrotrons operate near the cut-off frequency of the desired operating mode in the cavity. The frequency of the resulting $TE_{m,p}$ mode resonance is set to correspond to the desired operating frequency, which in turn must satisfy the cyclotron resonance condition. At the resonance frequency f the length of the middle section of the resonator is approximated by:

$$L_R \approx n \cdot \lambda_{\shortparallel}/2 \tag{1.4}$$

where $\lambda_{\shortparallel}$ is the wavelength in the resonator and n is the number of field maxima in the axial direction (the axial mode index). Usually it is taken as one in high power gyrotrons and it is neglected in the further contents of this thesis for the better readability. L_R is typically about 5-10 free space wavelengths λ_0 [Ker96].
The field amplitude in axial direction can be described by a complex function $f(z)$ (see chapter 2). Fig. 1.2b shows a typical $f(z)$ of an operating mode ($TE_{-34,19}$), calculated along the cavity having maximum at the straight portion of the interaction cavity. The electric field structure in the resonator is determined by the excited eigenmode in the cavity [JPT+06]. Fig. 1.3 shows some typical TE-eigenmodes. For a cylindrical waveguide cavity, the eigenmodes are given as $TE_{m,p}$, $TM_{m,p}$ -modes and, especially for a coaxial structure, the TEM-mode. Efficient energy exchange with $TM_{m,p}$ modes in the cylindrical waveguide structure is not possible as the

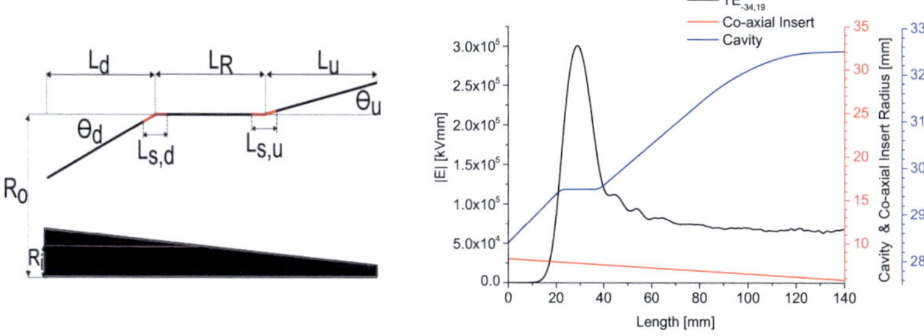

Figure 1.2: (a) Schematic diagram of a coaxial cavity with a tapered insert and (b) typical example of a longitudinal field profile inside the cavity.

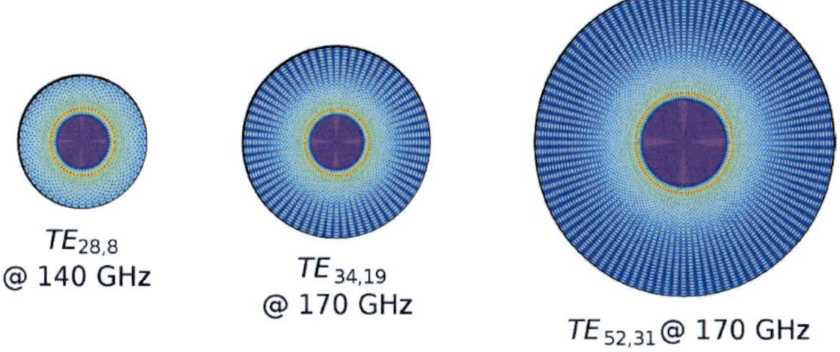

Figure 1.3: Typical eigenmodes used in gyrotrons (from [Ber11]).

transverse components E_φ and E_r vanish in the portion of the resonator close to the cut-off. As the TEM-mode of a coaxial cavity has no cut-off frequency it is not possible to have reflections towards the down taper section of the resonator which makes a resonance with a high quality factor impossible. It results, in typical gyrotron operation only $TE_{m,p}$ modes are excited as for these modes transverse components do not vanish and also the dimensions of the resonator have no limitation with respect to the wavelength of the radiation. It should be mentioned that different rotations of the mode relative to the helical motion of the electrons is reflected by different signs of the azimuthal index $\pm|m|$ ($TE_{-m,p}$ = co-rotating mode and $TE_{+m,p}$ = counter-rotating mode). The maximal coupling of $TE_{-m,p}$ modes to the electron beam is slightly higher which makes them preferable [Ber11].

In high power gyrotrons, non-linear tapers are used to connect the interaction region with the output waveguide system. The presence of a taper in a waveguide introduces unwanted parasitic modes. The main challenge in the design of a non-linear taper for a gyrotron is that it requires very high transmission (above 99%) with very low spurious mode generation. Due to high output power, even 1% of reflections can cause severe damage to the entire system. Gyrotron

output tapers should act as a nearly perfect match at the input port of the launcher section with suppressed spurious modes at the output port of the uptaper section of the cavity with a taper length as short as possible. Non-linear tapers with gradual change of the radius cause low power conversion to spurious modes. The methods used for the analysis and synthesis of gyrotron output tapers are given in detail in [KBT04].
After the beam-wave interaction inside the cavity, the electron beam and RF wave travels towards the quasi-optical (QO) output system. This quasi-optical mode converter consists of a waveguide antenna launching a wave beam onto several deflecting mirrors. In most high-power gyrotrons, the QO mode converter system consists of a waveguide launcher, a quasi-parabolic or quasi-elliptical mirror and two or three focussing mirrors. It is placed in between gyrotron interaction cavity and the RF output window. A typical and simple example of launcher, usually used in gyrotrons, is a Vlasov type launcher, which consists of a cylinder whose end opening is cut along a helix [HMH$^+$03]. By using a specific inner wall perturbation along this launcher, a mode mixture can be generated which consists of several neighbouring modes having different orders [BD04] and which forms a Gaussian like intensity distribution. A detailed description can be found in [JTP$^+$09, Fla12].
At the end of the QO mode converter of a gyrotron, the generated RF power is coupled to the transmission line through a dielectric vacuum RF window. It serves as a barrier between the vacuum side of the gyrotron and the output transmission line. It must withstand high power, mechanical and thermal stresses, be leak tight ($\approx 10^{-10}$Torr) and should have very low absorption. At the end of the QO mode converter of a gyrotron, the generated RF power is coupled to the transmission line through a dielectric vacuum RF window which is an important component and serves as a barrier between the vacuum side of the gyrotron and the output transmission line. It should withstand high power, mechanical and thermal stresses, be leak tight ($\approx 10^{-10}$Torr) and be should have very low absorption.
Finally after the separation of the electron beam from the RF wave, the spent electron beam travels towards the collector where a depressed collector system allows recovery of some of the residual energy from the electron beam and thus can enhance the overall device efficiency.

1.1.1 The electron cyclotron interaction

The electron cyclotron interaction was initially discovered as a quantum mechanical effect of the coherent emission and absorption of radiation, and later on it was designated as the gyrotron ECM ('electron cyclotron maser') or CRM ('cyclotron resonance maser') instability. It should be noted that in addition to ECM or CRM mechanism, a kinetic theory of interaction exists in which the entire electron beam is considered as an active medium [DL92].
The electron-cyclotron-interaction, which in this case is simply called gyrotron interaction, describes the mechanism of the energy exchange of the rotational motion of the electron beam and the RF wave. A brief review of the principles and theory of gyrotron and its components

are presented in the subsequent part of this chapter. A more complete and detailed description can be found in [KBT04, FDJ$^+$99, Lit08].
Due to the electron-cyclotron-interaction, electromagnetic energy is radiated by weakly relativistic electrons gyrating in an external magnetic field which is produced by a magnet. In this case, the effective frequency, which corresponds to the relativistic electron cyclotron frequency Ω_c, is given by [KBT04]:

$$\Omega_c = 2\pi f_c = \frac{eB}{m\gamma} \approx 2\pi \frac{28\text{GHz} \cdot B/T}{\gamma} \tag{1.5}$$

where γ (relativistic factor) is given by:

$$\gamma = \frac{1}{\sqrt{1-(v/c)^2}} = 1 + \frac{W_{kin}}{mc^2} \approx 1 + \frac{W_{kin}}{511keV} \tag{1.6}$$

where W_{kin} and v are the kinetic energy and velocity of the electron beam respectively. The weakly relativistic electrons gyrating in a strong magnetic field are radiating energy coherently due to bunching caused by the relativistic mass dependence of their gyration frequency. Bunching is achieved because, as an electron loses energy, its relativistic mass decreases and it thus gyrates faster (see Fig. 1.4). The strength of the magnetic field determines the radiation frequency. It should be mentioned that when ω_{RF} is exactly synchronous to the Ω_c some electrons are accelerated while some electrons are decelerated depending upon the position of the electron relative to the phase of the RF field. In the non-relativistic case the net energy transfer from the electron beam to the RF wave averages out to be zero. However, in the relativistic case due to the changing relativistic factor of each electron there is a non zero net energy transfer and as a result generation and enhancement of the RF field takes place. As a consequence of this relativistic effect, the frequency difference between RF wave and the electrons movement $|\omega_{RF}$-$\Omega_c|$ changes and thus the relative phase difference for every electron changes over time.
If we assume that $\omega_{RF}>\Omega_c$, the behaviour of electrons initially losing energy differs from that of electrons initially gaining energy. For electrons which are in the acceleration phase γ increases, thus Ω_c decreases and as a result the frequency difference with respect to the initial higher ω_{RF} increases. Over time, accelerated electrons are accumulated with an increasing phase with which they tend to leave this undesired phase position more quickly. In contrast, as indicated in Fig. 1.4, for decelerated electrons their Ω_c increases and thus the frequency difference with respect to ω_{RF} decreases, which means that these electrons tend to remain longer in their desired phase position relative to the RF-field. Thus when $\omega_{RF}>\Omega_c$, electrons remain longer in the phase relative to the RF field, in which they emit their energy, since the change in phase per oscillation period decreases. In summary, the particles accumulate in a particular relative phase position where electrons transfer energy to the RF field. In cylindrical cavity gyrotrons with radius R_0 the operating mode is chosen close to cut-off ($v_{ph} = \omega_{RF}/k_{\shortparallel} >> c$), v_{ph} is the wave phase

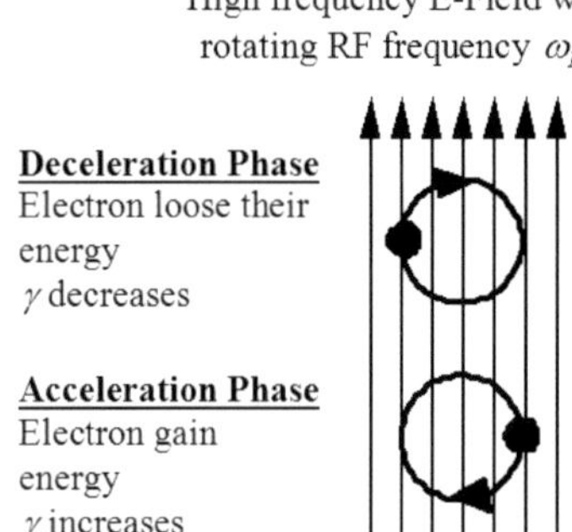

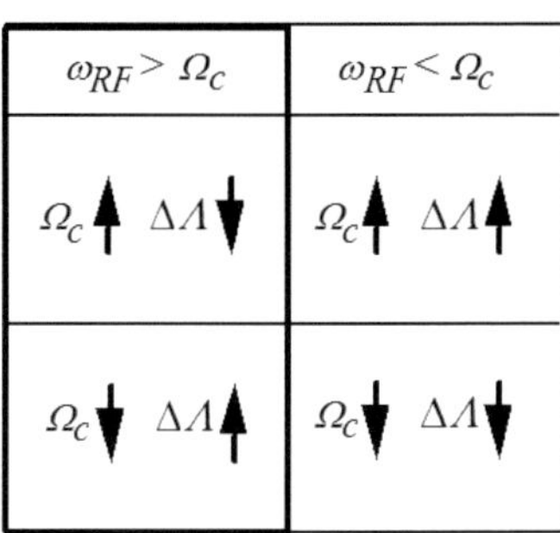

Figure 1.4: Principle of electron cyclotron interaction ([Thu13]).

velocity. To keep the bunched electron in the retarding phase, the frequency mismatch $|\omega_{RF}\text{-}\Omega_c|$ is chosen small but positive. The Doppler term $k_{\shortparallel}v_{\shortparallel}$ is of the order of the gain width and is small compared to the radiation frequency.

In the above description of the electron-cyclotron-interaction, it is required to consider the movement of the electrons in the direction of the flow of the electron beam. The electrons pass through the cavity with an axial velocity $v_{\shortparallel}$, within a certain transit time. Due to the presence of the electron axial velocity, the operating frequency is shifted by the Doppler term $k_{\shortparallel}v_{\shortparallel}$ where $v_{\shortparallel}$ is the axial wave number of the operating mode. The total wave number is given by:

$$k = \frac{2\pi}{\lambda} = \frac{2\pi f}{c} \tag{1.7}$$

with

$$k^2 = k_{\shortparallel}^2 + k_{\perp}^2 \tag{1.8}$$

To ensure that the phase of the electric field is synchronous to the rotational movement of the electrons, the phase difference over time needs to be

$$\frac{d}{dt}(\omega_{RF}t - k_{\shortparallel}v_{\shortparallel} - \Omega_c t) \geq 0 \tag{1.9}$$

Considering the Doppler shift, the resonance condition becomes

$$\omega_{RF} - k_{\shortparallel}v_{\shortparallel} \geq \Omega_c \tag{1.10}$$

Gyrotrons can also oscillate at higher cyclotron harmonics $s\Omega_c$. s is the harmonic number in which the RF field is inhomogeneous to the rotational movement of the electrons [Nus04]. Harmonic operation reduces the required magnetic field for a given frequency by the factor s.

As a result, the resonance condition becomes:

$$\omega_{RF} - k_{\shortparallel} v_{\shortparallel} \geq s\Omega_c \tag{1.11}$$

1.1.2 Dispersion diagram of a gyrotron

The operation of microwave tubes can be compactly presented with the help of *dispersion (Brillouin) diagrams* [Thu13]. Dispersion diagrams show the relation between a waveguide cavity mode

$$\omega_{RF}^2 = k_{\shortparallel}^2 c^2 + k_{\perp}^2 c^2 \tag{1.12}$$

and the fast cyclotron mode

$$\omega_{RF} - k_{\shortparallel} v_{\shortparallel} = \Omega_c \tag{1.13}$$

Here ω_{RF} is the RF frequency. The cyclotron resonance occurs at that point, for which the two frequencies obtained from equations (1.12) and (1.13) are equal. The typical dispersion diagram of a gyrotron oscillator is shown in Fig. 1.5 where equations (1.12) and (1.13) are plotted simultaneously i.e. the RF wave frequency is shown as a function of the axial wave number . In gyrotron oscillators, the measured efficiencies at higher harmonics ($s = 2$ & 3) are lower than those operating at the fundamental frequency ($s = 1$).

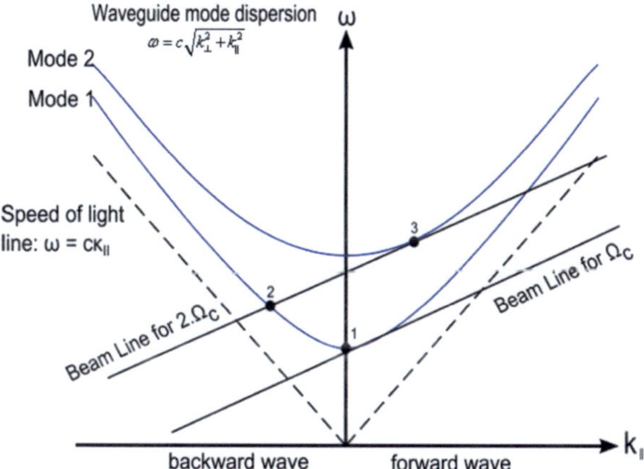

Figure 1.5: Dispersion diagram showing the region of interaction between the waveguide cavity modes, the beam, and beam harmonics.

Using a dispersion diagram, different operating points of the gyrotron interaction can be demonstrated by selecting points where the waveguide mode hyperbola curves touch or intersect with the beam line. In Fig. 1.5, point 1 corresponds to the classical gyrotron interaction with mode 1 near cut-off region at the fundamental electron cyclotron frequency Ω_c. The Doppler term is kept small, and the oscillating frequency is in first approximation equal to the cyclotron frequency. Point 2 in Fig. 1.5 represents a backward wave interaction with mode 1, since the corresponding $v_{\parallel}$ is negative where the waveguide dispersion hyperbola and the beam line of the second harmonic $2\Omega_c$ intersect. Point 3 corresponds to the interaction of a forward travelling mode 2 with the second harmonic of the electron cyclotron frequency. Point 3 represents the gyro-TWT type interaction. For numerical calculation of the gyrotron interaction more complicated models have to be considered. The corresponding theories can be found in [Ker96, Jel00].

1.1.3 Formalism of normalized variable

Generally the Brillouin diagram is used as a suitable way for the first estimation of the beam-wave interaction mechanism. However, for more detailed analysis of the gyrotron interaction it is necessary to consider detailed differential equations derived from Maxwell equations and the Lorentz force equation for the field profile and electron motion respectively. A ballistic theory can be used to estimate the trajectories of the annular electron beam inside the gyrotron cavity, electronic efficiency as well as impact of the calculated electric field on the annular electron beam.

Another possible way to describe the nonlinear theory of a gyrotron oscillator is the use of universal pendulum equations [DT86]. These generalized differential equations describe the evolution of the electron energy and their phase with respect to the RF field. The normalized variables method [KDS$^+$85] allows the evaluation of the efficiency of the electron-cyclotron-interaction for a broad mode spectrum. Summaries of the theory can be found in [IKP96, DT86, KT80]. [Bor91] gives an overview on different systems of dimensionless systems and variables. The method had verified success in several gyrotron experiments and is simple to adapt to various problems.

According to the normalized variable approach and with many simplifications and assumptions, the transverse energy u of a single particle can be given by [KDS$^+$85]:

$$\begin{aligned} \frac{du}{d\zeta} &= 2Ff(\zeta)\sqrt{1-u}\sin\theta \\ \frac{d\theta}{d\zeta} &= \Delta - \mu - 2\frac{Ff(\zeta)}{\sqrt{1-\mu}}\cos\theta \end{aligned} \tag{1.14}$$

The three normalized variables such as the field amplitude F, the interaction length μ and the detuning factor $\triangle$ can be written as:

$$
\begin{aligned}
F &= \frac{E_0 \beta_\perp^{-3}}{B_R} J_{m-1}(k_\perp R_e) \\
\mu &= \pi \left(\frac{\beta_\perp^2}{\beta_\shortparallel} \right) \left(\frac{\mathrm{L}_R}{\lambda_0} \right) \\
\triangle &= \frac{2}{\beta_\perp^2} \left(1 - \frac{f_c}{f} \right)
\end{aligned}
\tag{1.15}
$$

where $\beta_\perp$, $\beta_\shortparallel$, B_R, R_e, f_c, f, θ is the normalized electron velocity perpendicular to the magnetic field, the normalized electron velocity parallel to the magnetic field, the maximum magnetic field value in the gyrotron resonator, the electron guiding center radius, the relativistic electron cyclotron frequency, the frequency of the RF wave and the slowly varying gyro-phase of the electron respectively. With the effective cavity length ($\mathrm{L}_R = 2/k_\shortparallel$)) the normalized energy variable u as well as a axial position variable ζ is given by:

$$
\begin{aligned}
\mu &= \frac{2}{\beta_{\perp,in}^2} \left(1 - \frac{\gamma}{\gamma_{in}} \right) \\
\zeta &= \pi \left(\frac{\beta_\perp^2}{\beta_\shortparallel} \right) \left(\frac{z}{\lambda_0} \right)
\end{aligned}
\tag{1.16}
$$

Considering an axial Gaussian field profile $f(\zeta) = e^{-(2\zeta/\mu)^2}$, the electronic efficiency along with the transverse efficiency $\eta_\perp$ is given by:

$$
\begin{aligned}
\eta_{elec} &= \frac{\gamma_{in} - \gamma}{\gamma_{in} - 1} \\
&= \frac{\gamma_{in} \beta_{\perp,in}}{2(\gamma_{in} - 1)} \eta_\perp \\
\text{with} \qquad \eta_\perp &= \langle \mu(\zeta_{out}) \rangle_{\theta_{in}}
\end{aligned}
\tag{1.17}
$$

where $\eta_\perp$ is the average over an adequate number of electron input phase positions θ_{in}. The results of the numerical integration of equation (1.14) can be presented in the form of a single plot of the efficiency $\eta_\perp$ in the $F - \mu$ space (see Fig. 1.6). In this particular given example $F - \mu$ plot along with the Gaussian field profile, the transverse efficiency has it maximum value of 72% at $\mu = 19$ and $F = 0.13$.

Although the normalized variable system is more elegant and expresses the whole gyrotron interaction process with dimensionless as well as with sufficiently accurate parameter values, in the in-house developed KIT SELFT code these normalized variable approach is not used for the beam-wave interaction calculations due to the following two reasons: First these normalized variable system incorporates some more approximations than actually used in the SELFT code (e.g. velocity spread is not included), second the normalization factors usually include beam

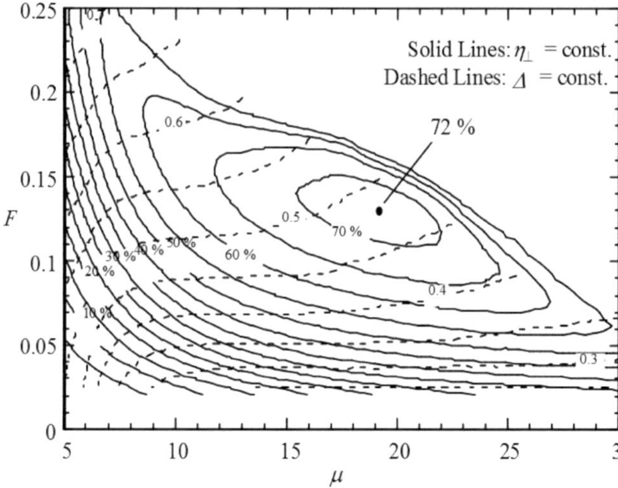

Figure 1.6: Transverse efficiency contour $\eta_\perp$ (solid line) as a function of the normalized field amplitude F and normalized effective interaction length μ for optimum detuning Δ (dashed line) [Ker96].

parameters which change in different situations during performing actual gyrotron experiments (e.g. during startup). The normalized variable approach is very powerful for basic studies but does not appear really appropriate for simulations codes which are set up for the design and experimental comparisons.

3D numerical methods like the finite difference method or the finite element method without any approximations, for field calculations, require large computational memory as well as very long computational time due to the large dimension of the resonator and uptaper as compared to the free-space wavelength of the propagating wave. Such 3D field calculation methods, therefore, are only suitable for problems of relatively low dimension.

1.2 Design considerations for CW gyrotron

Generally the design of high power gyrotrons involves trade offs among a number of design goals and constraints. In this section, some general design considerations particular for high power gyrotrons design in terms of the main technological as well as physical design constraints are discussed. Gyrotron design specifications and goals become more and more difficult if one wishes to work at very high frequencies, high power levels from long-pulse to CW operation, and with higher cyclotron harmonic number. The main design constraints for high power CW gyrotrons are shown in Table 1.1. The first four constraints given in Table 1.1 are related to electron beam physics whereas the emitter ring radius and the corresponding current density are technological constraints. The magnetic compression has a direct relation to the emitter radius.

Constraints	Limit (for CW operation)
TechnologicalConstraints	
Peak ohmic wall loading (ρ_{ohm})	$\leq 2.0\text{kW}/\text{cm}^2$
Emitter current density (J_E)	$< 3-5\text{A}/\text{cm}^2$
Electric field at the emitter (E_E)	$\leq 70\text{kV}/\text{cm}^2$
Emitter mean radius (R_E)	$\leq 5.5-8\text{cm}$
PhysicalConstraints	
Ratio of beam current to limiting current (I_b/I_L)	≤ 0.5
Magnetic compression ($b = B_R/B_E$)	< 50.0
Fresnel parameter (C_F)	$\geq 0.7-1.0$
Emitter/space charge current density (J_E/J_{SC})	$\geq 0.15-0.3$

Table 1.1: Main technological and physical constraints for CW gyrotron design [Pio93]

Cavity wall losses

The consideration of the cavity wall losses is important for long pulse to CW operation of high power gyrotrons. Requirements for cooling depend on the wall loading as well as the total losses [KR81, KKD+97, PBD+99]. Cavity wall losses are directly related to the quality factor of the resonator and the operating mode [Ker96]. The total output power P radiated from the type of cavity used in gyrotrons is composed of the radiated power P_{dif} as well as the power due to ohmic loss P_{ohm}, so accordingly the diffractive quality factor Q_{dif} as well as ohmic quality factor Q_{ohm} are defined by:

$$\begin{aligned} Q_{dif} &= \omega \frac{W}{P_{dif}} \\ Q_{ohm} &= \omega \frac{W}{P_{ohm}}, \qquad W = \text{Stored energy} \end{aligned} \tag{1.18}$$

Knowledge of the ohmic quality factor is essential for efficiency estimations [KBT04]. The actual quality factor is somewhat different due to the reflection ρ at the end of the cavity. For high power gyrotrons, the total quality factor Q is almost equivalent to the diffractive quality factor. The total quality factor for the resonator having a uniform straight part of the cavity is given by [FN88]:

$$\frac{1}{Q} = \frac{1}{Q_{dif}} + \frac{1}{Q_{ohm}} \tag{1.19a}$$

$$\text{with} \qquad Q_{dif} = 4\pi \frac{L_R^2}{\lambda_0^2}\left(\frac{1}{1-\rho}\right) \tag{1.19b}$$

where L_R is the effective length of the cavity. The ohmic quality factor of the cavity, excited by a $TE_{m,p}$ mode, is defined by:

$$Q_{ohm} = \frac{R_0}{\delta}\left(1 - \frac{m^2}{\chi_{m,p}^2}\right) \tag{1.20}$$

$\delta = \sqrt{\pi f \mu_0 \sigma}$ is the skin depth at the frequency f for the cavity radius R_0, σ is the conductivity, μ_0 is the absolute permeability, and $\chi_{m,p}$ is the eigenvalue of the $TE_{m,p}$ mode where m and p denote its azimuthal and radial index. Having the surface area of the cavity ($S = 2\pi R_0 L_R$), the average ohmic power density $\vec{\rho}_{ohm}$ loss is described by [VZO76, BW86]:

$$\vec{\rho}_{ohm} = \frac{P_{ohm}}{S} = \frac{Q_{dif} \cdot P_{dif}}{Q_{ohm} \cdot S} \approx \frac{2\sqrt{\pi}}{c^3\sqrt{\mu_0 \sigma}} \cdot \frac{f^{2.5} Q_{dif}}{L_R/\lambda_0 \cdot (\chi_{m,p}^2 - m^2)} P_{dif} \tag{1.21}$$

Equation (1.21) can be used for an estimation of the power loss by considering a Gaussian field profile. The maximum power dissipation density is higher while assuming a Gaussian field profile by a factor of ≈1.6 and on the other hand numerical calculations of the field profile in the resonators for high power gyrotron show typically a factor of ≈1.4. Thus a more accurate result can be obtained by performing a numerical evaluation of the resonator field profiles for high power gyrotrons.

Voltage depression and limiting current

Due to the presence of the space charge of the electron beam, a negative potential is created which screens the electrons partially from the accelerating voltage. The difference between the accelerating cathode voltage U_k and the corresponding beam voltage U_B i.e. $|\Delta U| = |U_k - U_B|$ describes the voltage drop ('voltage depression') within the hollow electron beam with respect to the cavity wall. The voltage depression for a conventional cavity is given by [DK81, KBT04]:

$$\Delta U = -\frac{I_B}{2\pi\varepsilon_0 v_{\shortparallel}} \text{In}\left(\frac{R_0}{R_e}\right) \approx -60\text{V}\frac{I_B/\text{A}}{\beta_{\shortparallel}} \text{In}\left(\frac{R_0}{R_e}\right) \tag{1.22}$$

Similarly the voltage depression for a coaxial cavity is defined as [CB93]:

$$\Delta U \approx -60\text{V}\frac{I_B/\text{A}}{\beta_{\shortparallel}} \text{In}\left(\frac{R_0}{R_e}\right)\left(\frac{\text{In}(R_e/R_i)}{\text{In}(R_0/R_i)}\right) \tag{1.23}$$

I_B is the beam current and R_i is the coaxial cavity insert radius. The upper limit for the allowed voltage depression is roughly estimated to be around 10% of the applied cathode voltage [KBT04]. The limiting current is the current for which the voltage depression becomes so large that the beam cannot propagate ($\beta_{\shortparallel} \rightarrow 0$ and mirroring of the electron beam occurs). The limiting current should be at least twice as large as the operating beam current [KBT04]. The limiting current for the conventional and coaxial cavities are given by [GC84, CB93]:

$$I_{lim} \approx 17070\text{A}\frac{\gamma_0(\beta_{\shortparallel 0}/\sqrt{3})^3}{2\text{In}(\frac{R_0}{R_i})} \quad \text{(for conventional cavity gyrotron)} \tag{1.24}$$

$$I_{lim} \approx 17070\text{A}\frac{\gamma_0(\beta_{\perp 0}/\sqrt{3})^3}{2\text{In}(\frac{R_0}{R_i})(\frac{\text{In}(R_e/R_i)}{\text{In}(R_0/R_i)})} \text{ (for coaxial cavity gyrotron)} \tag{1.25}$$

It should be mentioned that the voltage depression of a conventional cylindrical cavity is much larger than that of a coaxial cavity gyrotron and so accordingly the limiting current for the conventional case is much smaller than that of the coaxial gyrotron case. Because of the greater margin in the limiting current and the less voltage depression, the coaxial cavity gyrotron exhibits considerable advantage over the conventional gyrotron.

Fresnel Parameter

The Fresnel parameter describes how a particular mode will oscillate in the resonator [KBT04]. It also describes the losses due to diffraction in a quasi-optical resonator with a given length L_R (see Fig. 1.2a). In case of the gyrotron operation near the cut-off frequency of the operating mode in the resonator i.e. $(2\pi/\lambda) \approx (\chi_{m,p}/R_0)$, the expression for the Fresnel parameter (C_F) is defined by [KBT04]:

$$C_F \cong \frac{\pi}{4} \cdot \frac{(\frac{L_R}{\lambda})^2}{\sqrt{\chi_{m,p}^2 - m^2}} \tag{1.26}$$

C_F is a measure for the deformation of the phase front in a tapered waveguide due to diffraction. [GFG+81] shows that when C_F<1 the transverse as well as the longitudinal structure of the RF field is not sufficient enough for onset of a particular $\text{TE}_{m,p}$ mode in the gyrotron cavity. Using C_F>1 a single-mode description is appropriate.

1.3 State-of-the-art of static and dynamic after cavity interaction

Megawatt and multi-megawatt gyrotrons are under careful investigation by several institutions worldwide [RPD+07, HGA+09, KST+08, DLM+08]. As a result, new effects are observed where it is assumed that some spurious oscillations are excited by the electron beam after the interaction cavity. One of those effects is called 'After Cavity Interaction (ACI)'which takes place in the uptaper section after the interaction zone, if proper synchronization conditions (1.10) for a beam-wave interaction of the gyrotron type are fulfilled [SN09]. In most publications [CSS+07, ZM04, SNA10], the possible impact of ACI is described as a stationary reduction in the overall efficiency caused by direct energy exchange after the cavity between the electron beam and the RF field where RF energy is transferred back to the electron beam leading to lower efficiency. Due to this, the energy distribution of the spent electron beam after ACI is less suitable for energy recovery through means of a depressed collector. This phenomenon is described as static ACI [ZM04]. While the above effect exists and may have a deteriorating influence on the operation of a gyrotron, other kinds of ACI have also been found in recent theoretical and experimental investigations on possible undesired oscillations and interactions in a gyrotron which is known as dynamic ACI [KAC+10]. Dynamic ACI is based on the same mechanism as static ACI, but it behaves differently since further RF waves are generated: The interaction after the cavity leads to a dynamic amplitude modulation, either with the main mode or with different modes at lower frequency. Dynamic ACI modulates the output power and limits the stability region of the operating mode thereby inhibiting operation at the desired output power. On the contrary, dynamic ACI which takes place further away from the cavity interaction region seems to be no more than a minor side effect of a usual gyrotron operation, since the cavity oscillation remains undisturbed in such cases. However parasitic modes at lower frequencies will increase the stray radiation level in the tube. To understand dynamic ACI both experimental and theoretical investigations are carried out [CSS+07, SCG+11, CKA+12]. A first indication of limitations on output power was found in experiments at KIT Germany, when the frequency step tunable gyrotron did not deliver the desired MW power, after the operating mode was changed [TAB+03]. In the first version this tube only reached 0.6 MW, and the corresponding simulation indicated that the reason was the onset of dynamic ACI at higher acceleration voltages. It should be mentioned that in these experiments, an oscillation operating in a different frequency was found [KSF+09].

1.4 Statement of the problem addressed

In recent years, spurious oscillations like dynamic ACI in high power gyrotron gained a special attention as an influencing factor on the overall efficiency as well as device performance. Dynamic ACI is currently under extensive analysis regarding both the experimental findings and the capability of the present interaction models and simulating codes to describe dynamic ACI as close as possible [KAC+10, SCG+11, CSA+11].
Generally, the interaction of the electron beam with a particular waveguide mode takes place in the region of the cavity where the oscillation frequency is close to the cut-off frequency of the waveguide. Normally in the region after the gyrotron interaction cavity (see Fig. 1.7), the wave becomes a travelling wave and the interaction stops. Therefore, in most gyrotrons, the device performance in terms of overall efficiency, output power and operating mode stability does not change after the electrons pass through the cavity. But in high-power gyrotrons due to high electron currents there is a great probability to exceeds the starting current for ACI. When electrons propagate along the decreasing magnetic field and simultaneously along the uptaper waveguide section at which the cut-off frequency of the operating mode decreases ($\omega_{cutt} = ck_\perp = c\chi_{m,p}/R_w(z)$ where $\chi_{m,p}$ is the eigenvalue for the $TE_{m,p}$ mode and R_w is the varying cavity wall radius) it may be possible that the cyclotron resonance condition is fulfilled for a certain mode and again beam-wave synchronization may take place. This fact is also depicted in Fig. 1.8 where the after cavity interaction point is drawn by the interception of two dashed lines: the parabolic dashed line shows the dispersion curve for a waveguide mode of a larger radius. The straight dashed line shows the beam line for electrons with axial velocity and cyclotron frequency different from those in the cavity region.

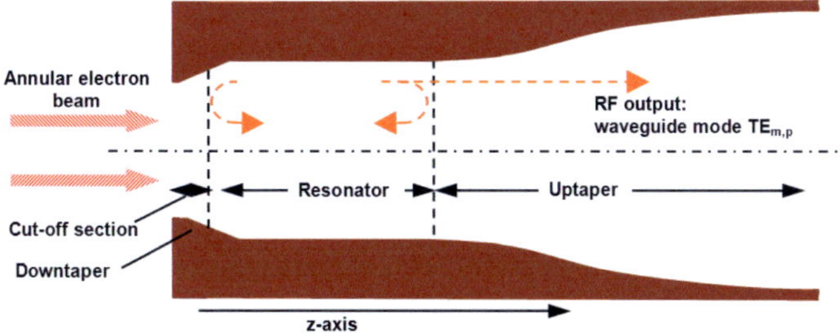

Figure 1.7: Gyrotron interaction cavity showing different sections.

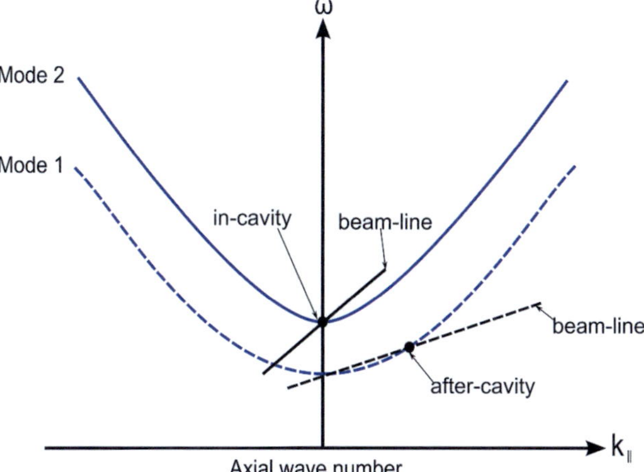

Figure 1.8: Dynamic ACI in Brillouin-Diagram shows resonance interaction in the cavity solid lines and after the cavity dashed lines.

At the Institute for Pulsed Power and Microwave Technology (IHM) of the Karlsruhe Institute of Technology (KIT), high power conventional as well as coaxial gyrotrons are under development in collaboration with several partners [Thu13], for example, high power gyrotrons with conventional cavity (e.g. 1 MW CW at 140 GHz for the Stellarator Wendelstein 7-X) and coaxial cavity (2 MW short pulse at 170 GHz for ITER) for fusion applications are being developed and verified experimentally. State-of-the-art high-power tubes are also being developed at the Institute of Applied Physics in Nizhny-Novgorod, Russia [DLM+08].

In the experiments performed at KIT Germany, frequencies of spurious oscillations are measured very precisely. For theoretical interpretations of the measured results with respect to dynamic ACI it is necessary to model the simulating tools as close to the reality as possible with adapting less assumptions in the simulating tool in order to support the measured results. In all previous investigations, spurious oscillations in the simulation results are treated as static or stationary ACI in which the overall efficiency is reduced by several percentages due to energy exchange between the electron beam and the outgoing travelling wave.

There are several European-Union (EU) codes which are based on the slow-time-scale approximation and uses different approaches for beam-wave interaction calculations like COAXIAL [Dum01], EURIDICE [AIP+08] and TWANG [ATA+11]. Comparing with these EU codes, the KIT developed SELFT code also utilizes direct time-dependent differential equations for beam-wave calculations which are directly derived from the fundamental Maxwell equations and the Lorentz force equation.

Using the SELFT code [Ker96] attempts were made to simulate the measured parasitic spectra. It has to be noted that SELFT originally was developed for the simulation of gyrotron interaction in the cavity and its vicinity where the magnetic field is considered to have a constant value. For calculation of ACI it is required to extend the simulation domain well into the uptaper section of the gyrotron cavity. Several assumptions in the code have only reduced validity with increasing distance from the cavity.

In the present work, comprehensive investigations has been carrier out with respect to dynamic ACI, in order to support the experimental results, by modifying the existing SELFT code in such a way that the time-dependent differential equations for beam-wave interaction calculations include a term containing a change in magnetic field. This means that SELFT calculations are also influenced by variation in the magnetic field along the cavity axis. In the present work it is the first time where the KIT SELFT code is used for extensive investigations of dynamic ACI present in the uptaper section of gyrotron interaction cavities and it will be shown that the tapered magnetic field and the contour of the waveguide have an important influence.

1.5 Content and structure of the thesis

This work is structured as follows:

In the present chapter 1, the main components and the operational principle of gyro-oscillators are introduced. In addition to that this chapter discusses the principal design considerations which are relevant for the design of high power gyrotrons. Subsequently this chapter also deals with the state-of-the-art of static as well as dynamic ACI investigations and also discusses the problems which are addressed in the present thesis.

Chapter 2 reviews the mathematical representation of the electron momentum and RF field profile equations.

In chapter 3, investigations of dynamic ACI phenomena with a uniform magnetic field profile, throughout the simulation domain, are addressed. In this chapter the adiabatic approximation is used to determine $u_{\parallel}(z)$ according to the varying magnetic field and to solve the resonance condition. Solutions for the resonance condition are then used to determine the approximate region(s) of the appearance of dynamic ACI in the simulation domain. In the subsequent part of this chapter the $u_{\parallel}(z)$ values, obtained according to adiabatic approximation, are used in the time-dependent self-consistent multi-mode simulations. KIT SELFT code RF spectra of the simulated electric field profiles are presented in order to determine the oscillation frequency of the main mode as well as parasitic/spurious modes.

In chapter 4 the discretized form of the differential equations, responsible for the beam-wave interaction calculations, and their numerical implementations in the KIT SELFT code which consider an axial inhomogeneous magnetic field term $dB_{R,z}/dz \neq 0$ are described. Necessary modifications to the existing code are described with respect to the slowly varying reference frequency calculated according to the varying magnetic field and the electron transit time influenced by the averaged electron parallel velocity taken over a number of macro-particles at each discrete point of the geometry.

In Chapter 5, convergence studies of the modified version of the SELFT code are presented with respect to the three numerical parameters: temporal step size, spatial step size as well as number of electrons required for azimuthal phase discretization. The modified differential equation required for the electron motion, solved by the Predictor-Corrector method is compared with the piecewise linearized analytical approach.

In chapter 6, dynamic ACI studies with the modified version of the SELFT code using four different gyrotron configurations (two conventional cavity gyrotrons and two coaxial gyrotrons) are presented. The simulated results for the W7-X 140 GHz $TE_{-28,8}$ mode gyrotron are compared with the available experimental results with respect to the parasitic/spurious oscillations present in the output section of the resonator. RF spectra of the simulated electric field profiles for each case are also presented. The simulated results presented in this chapter may be considered as a validation of all the proposed modifications made in the SELFT code, as described in chapter 4.

Chapter 7 summarizes the work towards the effect of the influence of the changing magnetic field and the contour of the cavity output taper on the appearance of dynamic ACI phenomena in the uptaper section, draws conclusion and gives an outlook addressing future issues and improvements that can be made in the existing modified SELFT code.

2 Physical-mathematical model implemented in SELFT

The core of the work described in this doctoral thesis consists of the numerical investigation of so-called ACI phenomena through the use of the SELFT code, developed at the Institute of Pulsed Power and Microwave Technology of KIT (former Forschungszentrum Karlsruhe) by S. Kern for his Ph.D. Thesis [Ker96]. In order to give a better understanding of the modifications in the SELFT code necessary for the calculation campaign performed in this work, and described in detail in Chapter 6, a short review of the equations implemented in the code is given in the following along with the necessary physical assumptions.

We anticipate here that, in the development of this work, it became clear that an extension of the existing numerical model was needed for a better description of the ACI phenomena. A term accounting for the changing magnetic field has to be explicitly introduced in the discretized equations (see chapter 4). This new item contributed to expand the range of applications of SELFT, whose fundamentals are given in this paragraph.

The implemented mathematical theory in the SELFT code for gyrotron interaction calculations, which can also be referred to as ballistic theory, is based on the theory which is described in [VZO+69, BMP+73, FRC+82, Bor93, CAS92]. It should be mentioned that SELFT can describe the beam-wave interaction well within the interaction cavity where the value of the magnetic field is taken as constant [Ker96]. The purpose of the present work is to study the behaviour of the beam-wave interaction in the uptaper from the interaction cavity (conventional cylindrical or coaxial) to the launcher where the magnetic field is no longer constant but varies with the axial.

The present section does not deal with the complete mathematical modelling of the gyrotron interaction which is actually used in the SELFT code (for more details see [Ker96]) but highlights only the required portion of the mathematical modelling which is directly influenced by the change of the magnetic field. In the SELFT code a time-dependent self-consistent approach has been adapted in the calculations which mean that the movement of a representative set of macro-particles (ensemble of electrons) is performed under the influence of the RF field in the resonator. So the change in magnetic field has a significant simultaneous influence on both the electron motion equation (Lorentz force equation) as well as the RF field profile equation. Following the expansion of theory to coaxial geometries as shown in [Ker96], this section gives a rough overview of the corresponding formulas required for calculation of gyrotron interaction influenced by non-uniform magnetic field.

2.1 RF field profile

The starting point for the derivation of the RF field equation are the Maxwell equations in the differential form:

$$
\begin{aligned}
rot\vec{\mathrm{H}} &= \nabla \times \vec{\mathrm{H}} = \vec{\mathrm{J}} + \frac{\partial \vec{\mathrm{D}}}{\partial t} \qquad & div\vec{\mathrm{B}} &= \nabla \cdot \vec{\mathrm{B}} = 0 \\
rot\vec{\mathrm{E}} &= \nabla \times \vec{\mathrm{E}} = -\frac{\partial \vec{\mathrm{B}}}{\partial t} \qquad & div\vec{\mathrm{D}} &= \nabla \cdot \vec{\mathrm{D}} = \vec{\eta}
\end{aligned}
\tag{2.1}
$$

Assuming homogeneous and isotropic materials, the time-harmonic ($e^{j\omega_k t}$) RF field is described by the following differential equation

$$
\Delta\vec{\mathrm{E}} - \frac{1}{c^2}\frac{\partial^2 \vec{\mathrm{E}}}{\partial t^2} = \mu_0 \frac{\partial \vec{\mathrm{J}}}{\partial t} + grad\ \frac{\vec{\eta}}{\varepsilon_0} \tag{2.2}
$$

The electric field $\vec{\mathrm{E}}$ can be splited into eigenwaves $\vec{\mathrm{e}}_k = (\vec{\mathrm{e}}_k^+ + \vec{\mathrm{e}}_k^-)/2$ in forward (-) and backward (+) direction at a given local resonator cross section. The index k represents the mode indices (m,p) of a $\mathrm{TE}_{m,p}$ mode and is used for better readability.

$$
\vec{\mathrm{E}} = \sum_k \left(\vec{f}_k^+(z,t) e^{jk_{\shortparallel,k} z} \vec{\mathrm{e}}_k^+ + \vec{f}_k^-(z,t) e^{-jk_{\shortparallel,k} z} \vec{\mathrm{e}}_k^- \right) e^{j\omega_k t} \tag{2.3}
$$

Here $\vec{\mathrm{e}}_k$ have the following characteristics

$$
\int_A \vec{\mathrm{e}}_{k'} \vec{\mathrm{e}}_k^* da = \delta_{k'k} \text{ A:Cross-section of the resonator} \tag{2.4a}
$$

$$
\text{and} \qquad \Delta_\perp \vec{\mathrm{e}}_k + \frac{\omega_{\perp,k}^2}{c^2} \vec{\mathrm{e}}_k = 0 \tag{2.4b}
$$

The transverse Laplace operator is given by $\Delta_\perp = \Delta|_{\partial/\partial z = 0}$. Here $\vec{f}_k^\pm$ describes the time-dependent complex field amplitudes. $\vec{f}_k^\pm$ can be splited into two components: the transverse component $\vec{f}_{\perp,k}$ and the longitudinal component $\vec{f}_{\shortparallel,k}$. For the sake of simplicity the transverse component $\vec{f}_{\perp,k}$ is represented by $\vec{f}_k$.

$$
\begin{aligned}
\vec{f}_k &= \left(\vec{f}_k^+(z,t) e^{jk_{\shortparallel} z} + \vec{f}_k^-(z,t) e^{-jk_{\shortparallel} z} \right)/2 \\
\vec{f}_{\shortparallel,k} &= \frac{k_{\perp,k}^2}{k_{\shortparallel,k}^2} \left(\vec{f}_k^+(z,t) e^{jk_{\shortparallel} z} - \vec{f}_k^-(z,t) e^{-jk_{\shortparallel} z} \right)/2
\end{aligned}
\tag{2.5}
$$

It has been mentioned before that efficient interaction in the gyrotron cavity is only possible with TE-modes because the transverse field components are not equal to zero. TE-modes do not have longitudinal electric field components $\vec{\mathrm{E}}_z = 0$ and can therefore be described solely with the transverse field portions $\vec{f}_k$.

Thus, the RF field $\vec{\mathrm{E}}$ in the gyrotron resonator can be described as:

$$\vec{\mathrm{E}} = \sum_k \vec{f}_k(z,t)\vec{\mathrm{e}}_k e^{j\omega_k t} \tag{2.6}$$

The angular frequency ω_k is chosen close to the cut-off frequency of the operating mode in the resonator. Thus the resulting equation for the RF field profile in the resonator without any simplifications, obtained by using equations (2.2), (2.5) and (2.6) and also by applying some mathematical operations, can be written as (for more details see [Ker96]):

$$\begin{aligned}
&\frac{\partial^2}{\partial z^2}(\vec{f}_k(z,t)) + \frac{\omega_k^2 - \omega_{\perp,k}^2}{c^2}\vec{f}_k(z,t) - \frac{1}{c^2}\frac{\partial^2}{\partial t^2}\vec{f}_k(z,t) - j\frac{2\omega_k}{c^2}\frac{\partial}{\partial t}\vec{f}_k(z,t) \\
&+\sum_k \left[\left(2\frac{\partial}{\partial z}(\vec{f}_k(z,t)) \int_A \frac{\partial}{\partial z}(\vec{\mathrm{e}}_{k'}) \cdot \vec{\mathrm{e}}_k^* da + \vec{f}_k(z,t) \int_A \frac{\partial^2}{\partial z^2}(\vec{\mathrm{e}}_{k'}) \cdot \vec{\mathrm{e}}_k^* da\right) e^{j\omega_k t}\right] e^{-j\omega_k t} \\
&= \mu_0 \left[\frac{\partial}{\partial t}\sum_j \left(\frac{\vec{\mathrm{u}}_j \vec{\mathrm{e}}_k^*}{u_{\shortparallel,j}} \cdot I_{B,j}\right)\right] \cdot e^{-j\omega_k t} + \int_A grad\frac{\vec{\eta}}{\varepsilon_0} \cdot \vec{\mathrm{e}}_k^* da \cdot e^{-j\omega_k t}
\end{aligned} \tag{2.7}$$

The first line of the RF field profile equation (2.7) contains the terms for the RF field profiles, the second line describes the coupling due to mode conversion whereas the third line describes the excitation and feedback term from the electron beam. This excitation term for the RF field profile is obtained by performing a summation over the number of macro-particles (ensemble of electrons) with the complex relativistic momentum $\vec{\mathrm{u}}_j$ (details of the description will be presented later on) times the beam current of each macro-particle $I_{B,j}$ and also with the longitudinal component of the relativistic momentum $u_{\shortparallel,j}$. In order to make SELFT simulations as fast as possible, equation (2.7) is implemented with several approximations, see [Ker96]. Out of several approximations, the three main simplifications are the negligence of mode conversion, of space charges and of fast time-dependent changes ($\partial^2/\partial t^2 = 0$). As a result equation (2.7) becomes

$$\begin{aligned}
\frac{\partial^2}{\partial z^2}(\vec{f}_k(z,t)) + \frac{\omega_k^2 - \omega_{\perp,k}^2}{c^2}\vec{f}_k(z,t) - j\frac{2\omega_k}{c^2}\frac{\partial}{\partial t}\vec{f}_k(z,t) \\
= \mu_0 \left[\frac{\partial}{\partial t}\sum_j \left(\frac{\vec{\mathrm{u}}_j \vec{\mathrm{e}}_k^*}{u_{\shortparallel,j}} .I_{B,j}\right)\right] .e^{-j\omega_k t} \\
= \vec{R}_k
\end{aligned} \tag{2.8}$$

In the above RF field profile equation equation (2.8) only the excitation term $\vec{R}_k$ depends explicitly on $\vec{\mathrm{e}}_k$ and the complex relativistic momentum of each particle which in turns also depends on the change in magnetic field along the axis of the resonator. Consideration of the dependency of $\vec{R}_k$ on the change in the magnetic field is essential in order to make investigations of dynamic ACI. The dependence of the $\vec{R}_k$ on the change in magnetic field will be described later. The fully coupled differential equation (2.8) is suitable for time-dependent self-consistent non-steady state simulation of each mode along with simultaneously solving the equation of motion for each particle. The equation for stationary state condition of the RF field profile is obtained by putting time derivative terms in equation (2.8) equal to zero i.e. $\partial/\partial t = 0$. In this case equation (2.8) can be written as:

$$\frac{\partial^2}{\partial z^2}(\vec{f}_k(z)) + \vec{k}^2_{\shortparallel,k} \cdot \vec{f}_k(z) = \mu_0 \left[\frac{\partial}{\partial t} \sum_j \left(\frac{\vec{\mathrm{u}}_j \vec{\mathrm{e}}^*_k}{u_{\shortparallel,j}} .I_{B,j} \right) \right] .e^{-j\omega_k t} \\ = \vec{R}_k \tag{2.9}$$

Stationary self-consistent single-mode calculations prove to be an important tool for cavity design. Finally, the cold cavity RF field profile can be obtained by putting the excitation term equal to zero [VZO+69] i.e. $\vec{R}_k = 0$ in equation (2.9).

$$\frac{\partial^2}{\partial z^2}(\vec{f}_k(z)) + \vec{k}^2_{\shortparallel,k} \cdot \vec{f}_k(z) = 0 \tag{2.10}$$

The cold cavity field profile provides initial field profile values along the resonator axis for the non-stationary time-dependent self-consistent simulations using the SELFT code.

2.2 Lorentz force equation

The aim of this derivation is to obtain time dependent differential equations for the slowly varying electron motion under the influence of the RF field. However again the intention is not to be exhaustive. The mathematical description for the electron movement are based on the equilibrium of the relativistic Lorentz- and the centrifugal-forces for the undisturbed rotating electrons. Theoretical approaches and summaries are gathered among many others in [FRC+82, Bor93]. The starting point is the relativistic Lorentz force equation which is given by:

$$\frac{d\gamma m\vec{\mathrm{v}}}{dt} = \Delta q(\vec{\mathrm{E}} + \vec{\mathrm{v}} \times \vec{\mathrm{B}}) \tag{2.11}$$

where Δq and m are the charge and mass of the electron and $\vec{\mathrm{v}}$ is it's velocity. B_R is the magnetic flux density in the resonator. The velocity of the electron can be substituted by a normalized velocity which is given by:

$$\vec{\mathrm{u}} = \frac{\gamma\vec{\mathrm{v}}}{c} \tag{2.12}$$

The normalized velocity of the particle can be split into the transverse velocity component $\mathrm{u}_\perp$ as well as the longitudinal velocity component $u_\shortparallel$ in the z-direction and from Fig. 2.1, it can be seen that $\mathrm{u}_\perp$ has also two components: one is in the radial direction i.e. radial component u_r

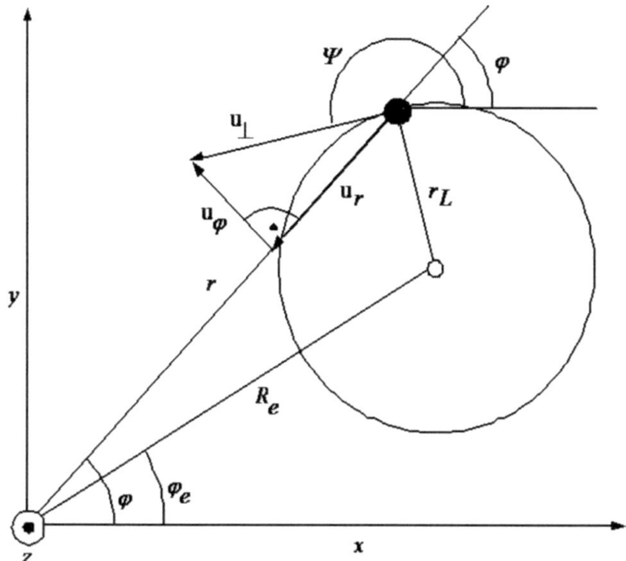

Figure 2.1: Schematic diagram showing transverse velocity components of a particle (bold vectors) rotating around the guiding center (R_e, φ_e) (circle). The angular position of the particle around the guiding center is described by ψ (Adapted from [Ker96] for illustration purpose).

and another is in the azimuthal direction i.e. azimuthal component u_φ. As a consequence the normalized velocity can be written as:

$$\begin{aligned} \mathrm{u} &= \mathrm{u}_\perp + u_{\shortparallel} e_z = u_r e_r + u_\varphi e_\varphi + u_{\shortparallel} e_z \\ \text{with} \quad u_r &= u_\perp \cos(\psi - \varphi) \quad u_\varphi = u_\perp \cos(\psi - \varphi) \end{aligned} \tag{2.13}$$

Using equation (2.11) and considering $\Delta q/m = -e/m$, the Lorentz force equation can written as follows

$$\frac{d\mathrm{u}_\perp}{dt} = -\frac{e}{m}\left(\frac{\mathrm{E}_\perp}{c} + \frac{(u_\varphi B_{R,z} - u_{\shortparallel} B_{R,\varphi})\, e_r + (u_{\shortparallel} B_{R,r} - u_r B_{R,z})\, e_\varphi}{\gamma}\right) \tag{2.14a}$$

$$\frac{d\mathrm{u}_{\shortparallel}}{dt} = -\frac{e}{m}\left(\frac{E_z}{c} + \frac{\mathrm{u}_\perp \times \mathrm{B}_{R,\perp}}{\gamma}\right) \tag{2.14b}$$

After inserting the components of $\mathrm{u}_\perp$ (2.13) into equation (2.14a), one gets:

$$\begin{aligned} &\frac{d}{dt}\left(u_\perp \left(\cos(\psi - \varphi) e_r + \sin(\psi - \varphi) e_\varphi\right)\right) \\ &= -\frac{e}{m}\left(\frac{\mathrm{E}_\perp}{c} + \frac{(u_\perp \cos(\psi - \varphi) B_{R,z} - u_{\shortparallel} B_{R,\varphi})\, e_r + (u_{\shortparallel} B_{R,r} - u_\perp \cos(\psi - \varphi) B_{R,z})\, e_\varphi}{\gamma}\right) \end{aligned} \tag{2.15}$$

The phase angle ψ can be described as the summation of the rapidly varying portion $\Omega_f t$, with an average angular frequency Ω_f, which is very close to the cyclotron frequency, the slowly varying portion ('Slow Variable Phase') and a phase offset value ξ.

$$\psi = -\Lambda(t) + \Omega_f t + \xi \tag{2.16}$$

In order to describe the bunching mechanism and the energy exchange with the RF field, it is necessary to calculate the slow phase variable $\Lambda(t)$ and the amplitudes of the velocity components.
Solving the time derivative portion i.e. left hand side of equation (2.15) using the time derivative of the product of sine and cosine functions having arguments $(\psi - \varphi)$ along with the time derivation of the eigenvectors and also using the time derivative of equation (2.16), the differential equation for the transverse velocity component can be written as:

$$\begin{aligned}
&\left(\cos(\psi-\varphi)e_r + \sin(\psi-\varphi)e_\varphi\right)\frac{du_\perp}{dt} \\
&+u_\perp\left(-\frac{d\Lambda}{dt}+\Omega_f\right)\left(-\sin(\psi-\varphi)e_r + \cos(\psi-\varphi)e_\varphi\right) \\
&= -\frac{e}{m}\left(\frac{\mathrm{E}_\perp}{c} + \frac{\left(u_\perp\cos(\psi-\varphi)B_{R,z} - u_{\shortparallel}B_{R,\varphi}\right)e_r + \left(u_{\shortparallel}B_{R,r} - u_\perp\cos(\psi-\varphi)B_{R,z}\right)e_\varphi}{\gamma}\right)
\end{aligned} \tag{2.17}$$

Decomposing equation (2.17) into the radial e_r as well as azimuthal e_φ eigenvector components and also by performing multiplication with a $\sin(\psi-\varphi)$ or $\cos(\psi-\varphi)$ term, addition and subtraction on the decomposed equations, the time derivative differential equation for the transverse component $u_\perp$ i.e. $du_\perp/dt$ as well as the time derivative of the slow phase variable i.e. $d\lambda/dt$ can be obtained (for more details see [Ker96]). The differential equation for the transverse electron movement can be written after introduction of the complex transverse momentum $\vec{p}$ [Ker96, Bor93].

$$\vec{p} = u_\perp e^{-j\Lambda(t)} \tag{2.18}$$

Applying the product rule of derivation to equation (2.18), the time derivative of the transverse electron momentum is given by:

$$\begin{aligned}
\frac{d}{dt}\left(\vec{p}(t)\right) &= \frac{d}{dt}\left(u_\perp(t)e^{-j\Lambda(t)}\right) \\
&= e^{-j\Lambda(t)}\left(\frac{d}{dt}u_\perp(t) - ju_\perp(t)\frac{d}{dt}\Lambda(t)\right)
\end{aligned} \tag{2.19}$$

Inserting the values of $du_\perp/dt$ and $d\lambda/dt$ into equation (2.19), the differential equation for the transverse electron momentum can be written as:

$$\begin{aligned}\frac{d}{dt}(\vec{p}) = &-\frac{e}{m}\left(\frac{E_r}{c}+j\frac{E_\varphi}{c}-\frac{u_{\shortparallel}(B_{R,\varphi}-jB_{R,r})}{\gamma}\right)e^{-j(\Omega_f t+\xi-\varphi)}\\ &+j\left(\frac{e}{m}\frac{B_{R,z}}{\gamma}-\Omega_f\right)\vec{p}\end{aligned} \tag{2.20}$$

Using vector components the longitudinal momentum from equation (2.14b) can be written as:

$$\frac{du_{\shortparallel}}{dt} = -\frac{e}{m}\left(\frac{E_z}{c}+u_\perp\frac{\cos(\psi-\varphi)B_{R,\varphi}-\sin(\psi-\varphi)B_{R,r}}{\gamma}\right) \tag{2.21}$$

Equations (2.20) and (2.21) are the differential equations of the motion of a single particle in the interaction cavity without any simplifications [Ker96, GA82]. The position of each particle is determined by calculating the components of $\vec{p}$ and $u_{\shortparallel}$, and then by integrating both equations (2.20 and 2.21) along the whole length of the interaction cavity. Equations (2.20 and 2.21) can be simplified using several assumptions. First it was assumed that all fast time varying magnetic field components are neglected because gyrotron operation takes place efficiently only with $\mathrm{TE}_{m,p}$ modes ($\vec{E}_z = 0$) close to the cut-off frequency and the transversal magnetic field components are very small in comparison to the external magnetic field [Ker96]. Also the coordinate system is chosen such that the magnetic field has no φ component i.e. $B_{R,\varphi} = 0$, this assumes rotational symmetry of the magnetic field. Another simplification is that the static magnetic field has only z-components. Thus in the SELFT calculations only the transverse RF electric field and external static magnetic field are used. The external static magnetic field in the resonator B_R can be decomposed into radial as well as axial direction which is given by:

$$\mathrm{B}_R = B_{R,r}e_r + B_{R,z}e_z \tag{2.22}$$

Using Maxwell's equation ($\nabla\cdot\vec{\mathrm{B}} = 0$) we get:

$$\frac{\partial(rB_{R,r})}{\partial r} = -r\frac{\partial B_{R,z}}{\partial z} \tag{2.23}$$

Integrating equation (2.23) by using near-axis approximation i.e. $B_{R,z}(z,r) \approx B_{R,z}(z,0)$ and the boundary condition $B_{R,r}(z,0) = 0$, the value of $B_{R,r}(z,r)$ can be defined by:

$$B_{R,r} = -\frac{r}{2}\frac{dB_{R,r}}{dz} \tag{2.24}$$

From Fig. 2.1, r (radius in cylindrical coordinate system) is determined from the movement of the electron and also by it's rotation with the *Larmorradius* ($r_L = v_\perp/\Omega_c$) around the guiding center whose position is determined by the radius R_e and the phase φ_e as shown in Fig. 2.1.

Thus from Fig. 2.1 it follows that [Ker96]

$$
\begin{aligned}
r\sin(\psi-\varphi) &= r_L + R_e\sin((\psi-\varphi_e)) \\
r\cos(\psi-\varphi) &= R_e\cos(\psi-\varphi_e)
\end{aligned}
\tag{2.25}
$$

In the electron momentum differential equations, the relativistic $\Omega_c(z,\gamma)$ as well as non-relativistic $\Omega_0(z)$ cyclotron frequencies are used which are given by:

$$
\frac{e}{m}\frac{B_{R,z}(z)}{\gamma} = \Omega_c(z,\gamma) = \frac{\Omega_0(z)}{\gamma}
\tag{2.26}
$$

Using equations (2.20, 2.21, 2.24, 2.25 and 2.26), the simplified equations for the electron movement with varying magnetic field can be written as:

$$
\begin{aligned}
\frac{d}{dt}(\vec{p}) + j(\Omega_f - \Omega_0/\gamma)\vec{p} - \left(\frac{cu_{\shortparallel}}{\gamma}\frac{1}{2B_{R,z}(z)}\frac{dB_{R,z}}{dz}\right)\vec{p} & \qquad (2.27a)\\
= -\frac{e}{m}(E_r + jE_\varphi)e^{-j(\Omega_f t+\xi-\varphi)} = \vec{a} &
\end{aligned}
$$

$$
\frac{du_{\shortparallel}}{dt} = -\left(\frac{cu_\perp^2}{\gamma}\frac{1}{2B_{R,z}(z)}\frac{dB_{R,z}}{dz}\right)
\tag{2.27b}
$$

The above differential equations are suitable for fast non-steady-state simulations in which the positions of the particles are still determined by performing numerical integration over time but in order to have fast and slowly vaying non-stationary simulations, as in the case of the SELFT code, it is necessary to convert differentiation over time into differentiation over the axial coordinate of the resonator by using the following transformation [Ker96]:

$$
\frac{d}{dt} = \frac{cu_{\shortparallel}}{\gamma}\frac{d}{dz}
\tag{2.28}
$$

For gyrotron cavities, the time evolution of the modes is much slower than the electron transit time of the electrons through the cavity. This means that the amplitude of the RF field profile remains constant during the electron transit time. Based on this assumption the coordinate transformation (2.28) is used to distinguish the fast and slow time scales. In this case the slow time scale is kept in time while the fast time scale is changed to a space coordinate. The physics remains the same with or without this transformation because both time scales are independent of each other. It is only a convenience to be able to plot electron bunching in space to be better able to visualize the electron position in space with each time step in the slow time scale.

Thus using equation (2.28), equation (2.27) with tapered magnetic field can be described as [Ker96]:

$$\frac{cu_{\shortparallel}}{\gamma}\frac{d}{dz}(\vec{p})+j\left(\Omega_f-\Omega_0/\gamma\right)\vec{p}-\left(\frac{cu_{\shortparallel}}{\gamma}\frac{1}{2B_{R,z}(z)}\frac{dB_{R,z}}{dz}\right)\vec{p} \tag{2.29a}$$

$$=-\frac{e}{m}\left(E_r+jE_\varphi\right)e^{-j\left(\Omega_f t+\xi-\varphi\right)}=\vec{a}$$

$$\frac{u_{\shortparallel}}{\gamma}\frac{du_{\shortparallel}}{dz}=-\left(\frac{u_\perp^2}{\gamma}\frac{1}{2B_{R,z}(z)}\frac{dB_{R,z}}{dz}\right) \tag{2.29b}$$

The above differential equations are used for the calculation of the electron motion, both in transverse as well as longitudinal direction, under the influence of the changing magnetic field. In order to have a self-consistent single-mode or multi-mode simulation, it is necessary to solve equation (2.29) in parallel to the RF field profile equation (2.8). The acceleration term $\vec{a}$ required for the electron motion in equation (2.29a) can be determined in terms of the electric field by using equation (2.6). The acceleration term for a single macro-particle $\vec{a}_j(r_j,\varphi_j)$ (see also Fig. 2.1) is given by:

$$\vec{a}_j\left(r_j,\varphi_j\right)=-\frac{e}{mc}\left[\mathrm{Re}\left(\sum_k \vec{f}_k\left(z,t\right)\left(\vec{\mathrm{e}}_{r,k}+j\vec{\mathrm{e}}_{\varphi,k}\right)e^{j\omega_k t}\right)e^{j\left(-\Omega_f t-\xi+\varphi\right)}\right]_{r_j,\varphi_j} \tag{2.30}$$

and the excitation term for the RF field profile from equation (2.8) is given by:

$$\vec{\mathrm{R}}_k=\mu_0\left[\frac{\partial}{\partial t}\sum_j\left(\frac{\vec{\mathrm{u}}_{\perp,j}\vec{\mathrm{e}}_k^*}{u_{\shortparallel,j}}\cdot I_{B,j}\right)\right]\cdot e^{-j\omega_k t} \tag{2.31}$$

A detailed derivation of the excitation term in terms of the current pulse is given in [FRC+82]. However, the detail derivation of the excitation term ($\vec{\mathrm{R}}_k$) required for the RF field profile used in SELFT code can be found in [Ker96]. Since in equation (2.31) the complex transverse velocity for each electron is considered which is given by:

$$\begin{aligned}\mathrm{u}_\perp&=u_\perp(\cos(\psi-\varphi)\mathrm{e}_r+\sin(\psi-\varphi)\mathrm{e}_\varphi)\\ \rightarrow\vec{\mathrm{u}}_\perp&=u_\perp(\mathrm{e}_r-j\mathrm{e}_\varphi)e^{\psi-\varphi}\end{aligned} \tag{2.32}$$

Thus equations (2.30) and (2.31) can be written as:

$$\vec{a}_j(r_j,\varphi_j)=-\frac{e}{2mc}\sum_k\vec{f}_k(z,t)(\vec{\mathrm{e}}_{r,k}+j\vec{\mathrm{e}}_{\varphi,k})e^{j((\omega_k-\Omega_f)t-\xi+\varphi_j)} \tag{2.33a}$$

$$\vec{R}_k=\mu_0\left[\frac{\partial}{\partial t}\sum_j\frac{I_{B,j}\cdot u_{\perp,j}}{u_{\shortparallel,j}}\left((\vec{\mathrm{e}}_{r,k}+j\vec{\mathrm{e}}_{\varphi,k})e^{-j(\psi_j-\varphi_j)}\right)^*\right]\cdot e^{-j\omega_k t} \tag{2.33b}$$

The excitation term $\vec{R}_k$ can also be processed further utilizing the product rule for differentiation and by using equation (2.16): $\psi = -\Lambda(t) + \Omega_f t + \xi$ and equation (2.18): $\vec{p} = u_\perp e^{-j\Lambda(t)}$. The transformed equation is given by:

$$\vec{R}_k = \mu_0 \sum_j \left[I_{B,j} \frac{\partial}{\partial t} \left(\frac{\vec{p}_j}{u_{\parallel,j}} \cdot \left[\left(\vec{e}_{r,k} + j\vec{e}_{\varphi,k} \right) e^{j((\omega_k - \Omega_f)t - \xi + \varphi_j)} \right]^* \right) + j\omega_k \left(\frac{\vec{p}_j}{u_{\parallel,j}} \cdot \left[\left(\vec{e}_{r,k} + j\vec{e}_{\varphi,k} \right) e^{j((\omega_k - \Omega_f)t - \xi + \varphi_j)} \right]^* \right) \right] \tag{2.34}$$

A comparison between equation (2.33a) and equation (2.34) show that both equations have a common factor which is given by [Ker96]:

$$(\vec{e}_{r,k} + j\vec{e}_{\phi,k}) e^{j((\omega_k - \Omega_f)t - \xi + \varphi_j)} \tag{2.35}$$

The eigenvectors for the RF field profile $(\vec{e}_{r,k}, \vec{e}_{\varphi,k})$ of a coaxial cavity gyrotron configuration can be described in terms of Bessel functions of the first kind and Neumann functions (Bessel function of the second kind) and are given by [Ker96]:

$$\begin{aligned} \vec{e}_{r,k} &= -\sum_m C_k \vec{A}_{m,k} \frac{jm}{r} (J_m(k_{\perp,k} r) + \vec{B}_{m,k} \cdot N_m(k_{\perp,k} r)) \cdot e^{jm\varphi} \\ \vec{e}_{\varphi,k} &= -\sum_m C_k \vec{A}_{m,k} k_{\perp,k} (J'_m(k_{\perp,k} r) + \vec{B}_{m,k} \cdot N'_m(k_{\perp,k} r)) \cdot e^{jm\varphi} \\ \text{with} \quad C_k &= \frac{1}{\sqrt{\pi(\chi_{m,p}^2 - m^2) J_m^2(\chi_{m,p})}} \quad \text{(Normalizing Factor)} \end{aligned} \tag{2.36}$$

Similarly, for the conventional cavity gyrotron configuration, the term $\vec{B}_{m,k}$ is considered to be zero and thus equation (2.36) can be written as:

$$\begin{aligned} \vec{e}_{r,k} &= -\sum_m C_k \vec{A}_{m,k} \frac{jm}{r} (J_m(k_{\perp,k} r)) \cdot e^{jm\varphi} \\ \vec{e}_{\varphi,k} &= -\sum_m C_k \vec{A}_{m,k} k_{\perp,k} (J'_m(k_{\perp,k} r)) \cdot e^{jm\varphi} \end{aligned} \tag{2.37}$$

where $\vec{B}_{m,k}$ is an intermediate constant and $\vec{A}_{m,k}$ is the amplitude factor. Inserting the values of the eigenvectors obtained from equation (2.36) and applying the additive property of Bessel functions, equation (2.35) can be described as:

$$\begin{aligned} &(\vec{e}_{r,k} + j\vec{e}_{\phi,k}) e^{j((\omega_k - \Omega_f)t - \xi + \varphi_j)} \\ &= -j \sum_m C_k \vec{A}_{m,k} k_{\perp,k} (J_{m+1}(k_{\perp,k} r) \\ &\quad + \vec{B}_{m,k} \cdot N_{m+1}(k_{\perp,k} r)) \cdot e^{jm\varphi} \cdot e^{j((\omega_k - \Omega_f)t - \xi + \varphi_j)} \end{aligned} \tag{2.38}$$

Since gyrotrons operate under the resonant interaction between the beam of rotating electrons and electromagnetic waves at the cyclotron frequency or its harmonics it is necessary that the electrons see a synchronous RF field profile for efficient beam-wave interaction. As a consequence Graf's Theorem [AS64] is applied to equation (2.38) [Ker96]. Thus equation (2.38) can be described by:

$$\begin{aligned}&(\vec{e}_{r,k} + j\vec{e}_{\varphi,k})e^{j((\omega_k-\Omega_f)t-\xi+\varphi_j)}\\ &= \underbrace{C_k k_{\perp,k} J_{m+1}(k_{\perp,k}R_e)}_{G_{e,k}} \underbrace{J_{s-1}(k_{\perp,k}r_L)}_{A} e^{j((\omega_k-s\Omega_f)t+(s-1)\Lambda(t))} \cdot e^{jm\varphi_e}\end{aligned} \tag{2.39}$$

where $G_{e,k}$ is the coupling factor and s is the harmonic number. For a relativistic electron beam with beam energy less than 250 kV, $k_{\perp,k}r_L <<1$ (Estimated by Borie [Bor93]), the factor A can be defined as:

$$A|_{k_{\perp,k}r_L<<1} \approx \frac{1}{(s-1)!}\left(\frac{k_{\perp,k}r_L}{2}\right)^{s-1} \tag{2.40}$$

Thus by using the above equations equations (2.39 and 2.40) and the definition of the complex transverse momentum (2.18) and *Larmoradius* ($r_L = v_\perp/\Omega_c$), the corresponding expressions for the excitation and acceleration terms (equations 2.33a and 2.34 respectively) for the special case of first harmonic i.e. $s = 1$ can be described by equation (2.41) [Ker96]. It can be seen from equation (2.41b) that the excitation term for the RF field profile is also influenced by the change of the magnetic field and this dependency is important to be considered in order to study or simulate the occurrence of dynamic ACI phenomena in gyrotrons [CKA+12].

$$\vec{a}_j(r_j,\varphi_j) = -\frac{e}{2mc}\sum_k \vec{f}_k(z,t)G_{e,k,j}\cdot e^{j(\omega_k-\Omega_f)t}\cdot e^{jm\varphi_{e,j}} \tag{2.41a}$$

$$\begin{aligned}\vec{R}_k = \mu_0 \sum_j \frac{I_{B,j}}{u_{\shortparallel,j}} G_{e,k,j} e^{-jm\varphi_{e,j}} \cdot e^{-j(\omega_k-\Omega_f)t}\\ \cdot\left(j\Omega_0 + \left(\frac{\gamma_j^2-1}{u_{\shortparallel,j}^2}\right)\left(\frac{1}{2B_{R,z}(z)}\frac{dB_{R,z}}{dz}\right)\right)\frac{p_j'}{\gamma_j}\end{aligned} \tag{2.41b}$$

In [Ker96] computer programs for the numerical solution of these time-dependent self-consistent equations have been developed with the assumption of a constant magnetic field which is not sufficient enough to study dynamic ACI phenomena in the gyrotron uptaper section. For the studies in the present thesis the existing SELFT code [Ker96] has been modified (see chapter 4) using the formulas reviewed in this chapter. To summarize, within self-consistent calculations, the field equations and the equations for the motion of the electrons have been solved simultaneously including time dependence, changing magnetic field and multi-mode scenarios.

3 Investigation of Dynamic ACI with uniform magnetic field and adiabatic approximation in the SELFT code

In this chapter the results of simulations, using the unmodified version of the SELFT code are described which show parasitic/spurious oscillations in the output section of the gyrotron cavity that may be attributed to the occurrence of dynamic ACI. In this study results are generated along with the assumption of constant $B_{R,z}(z) = B_0$ (maximum magnetic field) as well as with the implementation of the adiabatic approximation in the case of varying magnetic field. Due to the consideration of the adiabatic approximation, the axial velocity of each macro-particle can be determined. It depends upon the non-uniform distribution of the magnetic field. The varying axial velocities of each electron have been used to solve the well know synchronous condition equation (1.9)

$$\omega_{RF} - k_{\shortparallel} v_{\shortparallel} \geq \Omega_c \tag{3.1}$$

in order to estimate the region of fulfillment of the cyclotron resonance interaction condition between the spent electron beam leaving from the cavity and the traveling electromagnetic wave propagating towards the output part of a tube.
It was mentioned that the ACI, no matter whether it is a dynamic or static, occurs far away from the interaction region (see chapter 1). The dynamic ACI having a lower oscillating frequency than the main cavity oscillation seems to be no more than a minor side effect of the usual gyrotron operation, since the cavity oscillation remains almost undisturbed in such cases. This type of undesired oscillations probably mainly concentrate in the uptaper section, downstream from the cavity, where the magnetic field is slightly lower than in the cavity region (straight portion). Therefore, it is possible that in a certain cross sectional area of a gyrotron cavity output section a simultaneous tapering of the magnetic field and the waveguide radius will result in the fulfillment of the cyclotron resonance condition (3.1) between the electron beam and the traveling wave. In the Doppler term $k_{\shortparallel} v_{\shortparallel}$ in equation (3.1), both the axial wave number $k_{\shortparallel}$ as well as the axial electron velocity $v_{\shortparallel}$ vary along the axial z-coordinate. Moreover in the right-hand side of equation (3.1) the electron cyclotron frequency Ω_c is proportional to the non-uniform distribution of the magnetic field profile. Fig. 3.1 shows the gyrotron cavity region where the possibility of the fulfillment of the synchronous condition after the cavity is very high (shown by a dark ellipse) due the simultaneous variation of both the waveguide radius and magnetic field. Previously in the unmodified version of the SELFT code it has been assumed that the axial velocities of all electrons remain constant.

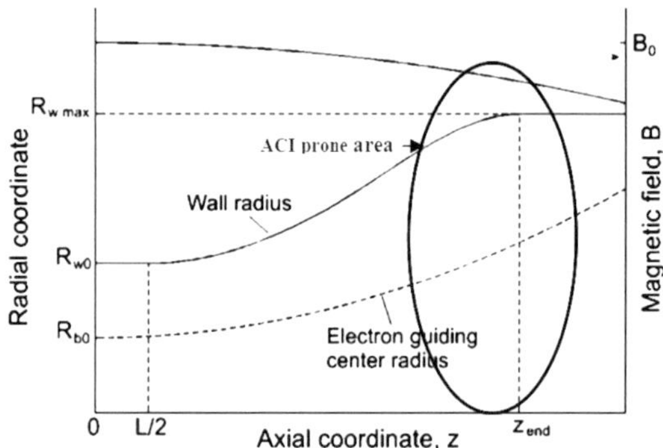

Figure 3.1: Schematic representation of axial profiles of the magnetic field, waveguide wall, and electron beam in the gyrotron after cavity region (adapted from [SN09] for illustration purpose).

In order to simulate as realistically as possible it is necessary to consider that electrons should have different axial velocity depending upon the non-uniform distribution of the magnetic field profile. It also requires that the transverse electron momentum should also depend upon the change in magnetic field (see chapter 4 for details). Thus before solving numerically the actual time-dependent differential equation (2.29b) for the electron axial velocity, initially adiabatic approximations are used to calculate the electron parallel velocity for all the electrons depending on the varying axial magnetic field profile. In this case still only the time-dependent differential equation (2.29a) is solved self-consistently with the RF field profile equation (2.8) in the unmodified version of the SELFT code assuming a uniform magnetic field profile. In reality, the magnetic field remains no longer constant and thus from Fig. 3.1, it is clear that in order to simulate spurious oscillations present in the output section of the cavity, it is necessary to include the regions after the cavity in the simulation domain i.e. the simulation domain should be extended up to the input of the launcher section.

In the initial numerical study, the first version of the $TE_{-22,8}$ mode frequency step tunable gyrotron of KIT was used in order to perform investigations on the parasitic/spurious oscillations present in the uptaper section since this tube did not deliver the desired MW output power, after the operating mode was changed [TAB+03]. The tube only reached 0.6 MW. Initially it was not clear that the presence of this additional frequency is the result of an additional cavity oscillation mechanism in a different mode, as proposed in [KSF+09], or if it represents the actual dynamic ACI, which has influence to the mode operating in the cavity [KAC+10].

3.1 Adiabatic Approximation

To describe the electron motion in crossed electric and magnetic fields, it is useful to separate velocity components into the gyrating motion and drift of the corresponding guiding centres. The approach, which is called adiabatic approximation, describes a very slow change of the field characteristics. It is valid, if the variations of the electric and magnetic fields are rather small at dimensions comparable with the electron trajectories [Pio93]. The conditions for adiabatic motion are given as [Che95, Siv65]:

$$z_L^2\left|\frac{\partial^2\vec{\mathrm{B}}}{\partial z^2}\right| \ll \left|\vec{\mathrm{B}}\right|; \qquad r_L^2\left|\frac{\partial^2\vec{\mathrm{B}}}{\partial R^2}\right| \ll \left|\vec{\mathrm{B}}\right| \tag{3.2a}$$

$$z_L^2\left|\frac{\partial^2\vec{\mathrm{E}}}{\partial z^2}\right| \ll \left|\vec{\mathrm{E}}\right|; \qquad r_L^2\left|\frac{\partial^2\vec{\mathrm{E}}}{\partial R^2}\right| \ll \left|\vec{\mathrm{E}}\right| \tag{3.2b}$$

$$\frac{\left|\vec{\mathrm{E}}\times\vec{\mathrm{B}}\right|}{\left|\vec{\mathrm{B}}\right|^2} \ll c \tag{3.2c}$$

During one rotation along the circle with the *Larmorradius* ($r_L = v_\perp/\Omega_c$) the electron travels a distance z_L in axial direction. Since $\left|\vec{E}\times\vec{B}\right|$ is the pointing vector (energy flux), equation (3.2c) indicates that the energy gain over one Larmor period has to be small. Thus, within the validity of the adiabatic approximation, the energy gain in axial direction over one Larmor period must not be higher than the initial beam energy. It results that the quantity $P_\perp^2/\mathrm{B}$ is an invariant of motion i.e.

$$\frac{p_\perp^2}{\mathrm{B}} = \mathrm{Constant} \tag{3.3}$$

where $p_\perp$ is the total transverse electron momentum and B is the local magnetic flux density. In the region between the gyrotron electron gun and cavity i.e. in the interaction free region, due to the increase in magnetic field amplitude (without any electric field) and also according to equation (3.3), the transverse momentum $p_\perp$ as well as velocity $\beta_\perp$ (transverse velocity $v_\perp$ normalized to the velocity of the light c) increases and as a consequence of conservation of beam energy, the axial momentum p_z as well as velocity $\beta_\parallel$ (axial velocity $v_\parallel$ normalized to the velocity of the light c) decreases thus it may be possible that the transverse momentum may become equal to the total momentum and simultaneously axial momentum becomes zero and thus at this condition electrons change their direction of propagation and become reflected electrons. Like in Busch's theorem i.e. for any two arbitrary positions (a) and (b) along the beam path the ratio between the change of the angular momentum and the change in the magnetic flux can be determined and similarly by using equation (3.3) it is possible to describe the relation for the transformation of the transverse momentum of the gyration motion at two arbitrary axial positions (a) and (b) which is given by:

$$p_\perp(b) = p_\perp(a)\left[\frac{\mathrm{B}(b)}{\mathrm{B}(a)}\right]^{1/2} \tag{3.4a}$$

$$\beta_\perp(b) = \beta_\perp(a)\frac{\gamma(a)}{\gamma(b)}\left[\frac{\mathrm{B}(b)}{\mathrm{B}(a)}\right]^{1/2} \tag{3.4b}$$

3.2 Adiabatic Approximation in the SELFT code

In the adiabatic approach we assume that at every point of the cavity we have γ for each macro-particle obtained from the beam-wave interaction calculations. It is also assumed that the transverse velocity components of each macro-particle at the entrance of the cavity ($\beta_{\perp,in}$) are known. Before the implementation of the proposed modifications (see chapter 4) in the actual differential equations (2.8 and 2.29) of the unmodified version of the SELFT code, with an aim to consider the effect of the non-uniform distribution of the magnetic field on the gyrotron beam-wave interaction calculations, it has been assumed that $\beta_{\shortparallel} \times \gamma = u_{\shortparallel}$ is constant for all macro-particles and also at every point of the gyrotron cavity [Ker96]. This assumption becomes incorrect when the magnetic field is not considered to have a constant value rather it varies along the axial coordinate of the cavity with a specific magnetic field profile. In the latter case, $\beta_{\shortparallel}$ can be calculated using the so called adiabatic constant i.e. $\beta_\perp^2/B_{R,z}$ = constant (this constant assumes that there was no beam-wave interaction). In order to determine the influence of the non-uniform distribution of the magnetic field profile on the beam-wave interaction phenomenon, initially the adiabatic approximation approach has been adapted where only initial values, i.e. values at the entrance of the cavity, for $u_{\perp,in}$ and $u_{\shortparallel,in}$ have been used for the calculations of $\beta_{\shortparallel}$. $\beta_\perp = u_\perp/\gamma$, for each value of axial coordinate (z), was then calculated within the validity of the adiabatic approximation using the expression given by equation (3.4b) and thus from the known values of $\beta_\perp$ and γ (calculated from the interaction calculations). Then $\beta_{\shortparallel} = u_{\shortparallel}/\gamma$ can be calculated at each z-coordinate using the analytical formula:

$$\begin{aligned} \gamma &= \sqrt{1+u_{\shortparallel}^2+u_\perp^2} \\ \gamma^2 &= 1+\gamma^2\beta_{\shortparallel}^2+\gamma^2\beta_\perp^2 \\ \beta_{\shortparallel} &= (1-\beta_\perp^2-1/\gamma^2)^{1/2} \end{aligned} \tag{3.5}$$

These locally changing values can then be used to calculate curves of the resonance frequencies according to equation (3.1) in order to determine the region(s) of interest along the simulation domain where fulfillment of the cyclotron resonance condition may occur due to the presence of both varying magnetic field profile as well as varying cavity radius. In these calculations the differential equation for electron motion (2.29a) is solved still without considering the axial change of the magnetic field profile explicitly in the differential equation, but the non-uniform magnetic field has been used in the calculations of the electron cyclotron frequency (Ω_c) and the changing beam radius.

As long as it was assumed that there is no beam-wave interaction in the calculations of $\beta_\perp$ (that means $\beta_\perp$ was calculated by using the adiabatic constant) it was straight forward to calculate $\beta_{||}$ for each macro-particle at each z value using equation (3.5). As a result the values $\beta_{||}$ can be used for initial estimation of intersection points between the mode's dispersion curve and the beam line in the Brillouin diagram by solving the resonance condition (3.1). On the other hand in order to include the actual beam-wave interaction in the calculation of $\beta_{||}$, the best way is to skip the approximations and to solve the actual differential equation for $\beta_{||}$ (2.29b). Then the values $\beta_{||}$ of can be used in the further calculations of $\beta_\perp$ which also includes the effect of the non-uniform distribution of the magnetic field (for more details see chapter 4). Similarly the calculated $\beta_\perp$ values can be used for the further estimation of the $\beta_{||}$ values.

3.3 Simulation results with unmodified version of the SELFT code and discussions

In Fig. 3.2, a typical single-mode calculation field profile featuring a dynamic ACI is shown. By including the regions after the cavity in the simulation domain, i.e. by considering the uptaper section, it has been observed that at the output section the field profile is varying (see Fig. 3.2) along the waveguide. This indicates that in this particular region, the beam-wave interaction condition is again fulfilled due to the presence of the tapered non-uniform magnetic field distribution that satisfies the synchronism condition with the outgoing radiation at some location in the uptaper section, where the varying radius is larger than the cavity radius. In order to confirm this unwanted phenomenon some numerical experiments have been carried out studying the influence of a uniform magnetic field distribution considered in the differential equations responsible for the calculations of the beam-wave interaction phenomenon. However, in reality the magnetic field is not uniform along the whole gyrotron resonator uptaper which has been considered in the subsequent part of this thesis in order to investigate the effect of non-uniform distribution of the magnetic field on the appearance of dynamic ACI in the uptaper section of the cavity geometry.

In the first numerical experiment, the second version of the $TE_{-22,8}$ mode 140 GHz step-frequency tunable gyrotron was taken into consideration in order to perform single-mode numerical investigations on the presence of dynamic ACI in the uptaper section of the geometry. In this beam-wave interaction calculations, the following beam parameters $V_{beam} = 80.0$ kV, $I_{beam} = 40.0$ A, $\alpha = 1.24$, $B_0 = 5.61$ T and $R_{beam} = 8.0$ mm have been used. Fig. 3.3 depicts the corresponding startup simulation, i.e. how the output power evolves during the time while the above electron beam parameters are achieved (linear increase). In Fig. 3.3, the vertical black line denotes the time ($\approx$860 ns) at which the electric field profile, shown in Fig. 3.2, is obtained and the final beam parameters have been achieved. Because of the modulation of the output power, as shown in Fig. 3.3, the display of power vs. time is not suitable for estimations of the modulation frequency of the desired operating mode as well as parasitic/spurious oscillating mode, if any. As a consequence it was necessary to plot the RF spectrum of the corresponding field profile so that the presence of any dynamic ACI in the uptaper section can be determined. As already mentioned in these numerical investigations, the simulated results have been obtained by using the unmodified version of the KIT SELFT code (i.e. solving only equation (2.29a)) in which a non-uniform distribution of the magnetic field has been taken into consideration only in the detuning factor (3.6a); this can be considered as a deviation only in the electron cyclotron resonance condition. This detuning factor in turn depends on the axial coordinate through the non-uniform distribution of the magnetic field profile i.e. the detuning factor in terms of change of magnetic field profile is given by (see equation 1.15):

$$\Delta = \frac{2}{\beta_{\perp,in}^2}\frac{\omega - \Omega_c}{\omega}, \quad \Omega_c = e/m\gamma B_{R,z}(z) \tag{3.6a}$$

and also in the changing beam radius i.e.

$$R_{beam}(z) = R_{beam}/(B_{R,z}(z)/B_0) \tag{3.6b}$$

where B_0 is the maximum value of the magnetic filed in the center of the interaction cavity.

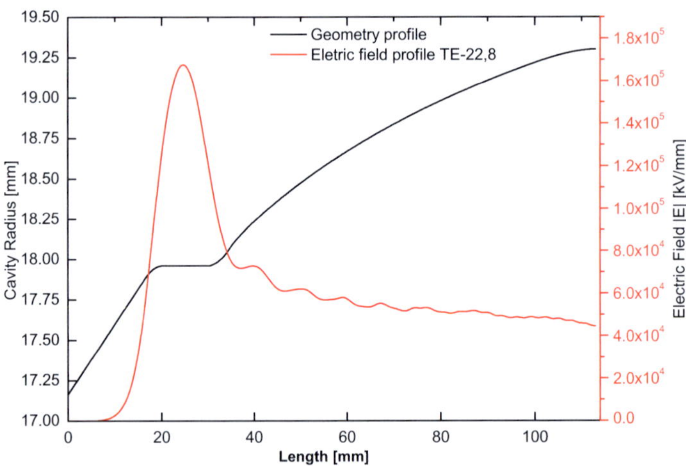

Figure 3.2: Geometry and field profile of an operating mode under influence of dynamic ACI (single-mode calculation).

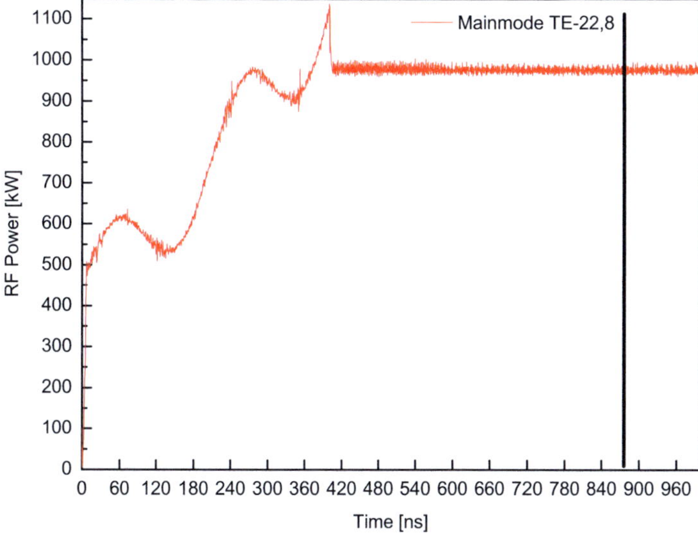

Figure 3.3: Startup simulation of RF power. The vertical black line denotes the time at which the field profile, shown in Fig. 3.2, is obtained (single-mode calculation).

In order to determine the reason of this modulating behavior of the RF field profile in the uptaper section of the geometry, initially the well known interaction condition (3.1) for electron cyclotron resonances has been solved to depict the regions of synchronism between RF wave field and electron beam. In order to investigate the dynamic ACI phenomenon due to the presence of the non-uniform distribution of the magnetic fields, it is more useful to solve equation (3.1) so that the value of ω_{RF} can be obtained along with the values of $k_{\parallel}$ which are calculated for the whole simulation domain. In this calculation the effect of the non-uniform distribution of the magnetic field has been taken into account according to the adiabatic approximation. These calculations also include the influence of the beam parameters through Ω_c calculations. Simultaneously the waveguide radius has an influence on these calculations through $k_{\parallel}$. The influence of the changing magnetic field as well as the waveguide radius are necessary to be included in the determination of the resonance curves in order to make a dynamic ACI study. In this single-mode (desired operating mode) simulation, still a constant value of $u_{\parallel}$ for each macro-particle has been considered in the differential equations used in the beam-wave interaction calculations and as a result a change in the magnetic field profile has no direct influence on the equations governing electron motion as well as RF field profile calculations. Solving equation (3.1) for ω_{RF} leads to a quadratic equation since $k_{\shortparallel}$ is also a function of ω_{RF} which is given by:

$$k_{\shortparallel} = \sqrt{k_0^2 - k_\perp^2}, \quad k_0 = \omega_{RF}/c, \quad k_\perp = \chi_{m,p}/R_0 \tag{3.7}$$

where k_0 is the free space wave number and c is the velocity of light. The result of possible interaction frequencies are shown in Fig. 3.4 along the z-coordinate of the cavity geometry (see Fig. 3.2) using the beam parameters at the end of the startup simulation shown in Fig. 3.3.

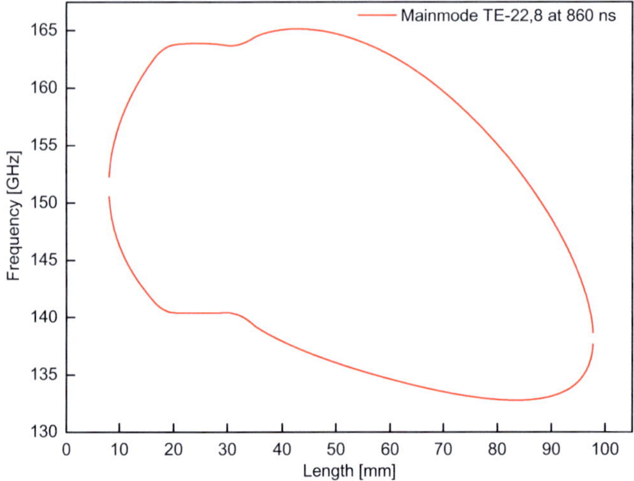

Figure 3.4: Frequencies that fulfill the interaction condition (3.1) along the cavity and uptaper geometry (see Fig. 1.7) at the end of the startup (see Fig. 3.3).

The two possible solutions shown in Fig. 3.4 (upper and lower curves) correspond to the two crossing points of a mode's hyperbola with the beam line (gyrotron and gyro-travelling wave interaction, see chapter 1) in the Brillouin-Diagram. Longer straight sections with nearly constant wave frequency ω_{RF} vs z. can be considered as suitable for a considerable net energy exchange between RF wave and electron beam. These possible curves can be used to depict the possible regions of interactions where the synchronization condition is valid along the length of the resonator and output waveguide. One can observe from Fig. 3.4 that there are some flat responses which can be seen as the region where the beam-wave interaction condition is fulfilled due to having a constant wave frequency along the z-coordinate: one region is $20.0mm \leq z \leq 30.0mm$ (cavity region) where the synchronization condition exists at

$\approx$140 GHz (desired frequency) whereas at a different location $75.0mm \leq z \leq 85.0mm$ another synchronization condition exits around 134 GHz which may be treated as undesired oscillation suspected to be related to dynamic ACI. If this is the excitation of a $TE_{-22,8}$ mode backward wave the output power modulation feature can be explained as beat structure of a 140 GHz forward wave and a 134 GHz backward wave. In order to plot solutions of equation (3.1) (see Fig. 3.4), here the axial velocity of the single particle is considered since it was assumed that all particles have the same axial velocity. It should be mentioned that the constant axial velocity of all macro-particles is due to the consideration of the uniform magnetic field profile throughout the simulation domain (see equation (2.29b), if $B_{R,z}(z) = B_0$ uniform magnetic field then $u_{\shortparallel} =$ constant for all macro-particles) and hence it is sufficient to consider a single particle in order to plot the resonance frequency curve, whereas in reality each particle has different axial velocity. One can also observe from Fig. 3.4 that another synchronization condition exists at the same desired interaction region ($20.0mm \leq z \leq 30.0mm$) as for the 140 GHz but with a frequency around 164 GHz which is far from the cut-off frequency designed for the chosen gyrotron resonator. Since a gyrotron operates near the cut-off frequency, this frequency corresponding to a gyro-TWT interaction and was not taken into consideration for the dynamic ACI investigations (see explanation of Fig. 3.9). Because of the modulation of the output power (see Fig. 3.3) and also as it has been mentioned earlier because the display of power vs. time is not suitable for estimation of the modulation frequency in order to have a closer observation, the RF spectrum of the corresponding varying electric field profile has been plotted in Fig. 3.5. The RF spectrum at the end of the simulation domain (see Fig. 3.43.2) indicates that, in addition to the desired operating 140 GHz oscillation of the cavity, another parasitic oscillation at $\approx$134.72 GHz may also exists. This parasitic oscillation in the cavity output section has been observed due to the consideration of the longer geometry in the simulation domain where the radial eigenvalue ($\chi_{m,p} \cdot R_0$) changes due to the change in cavity radius R_0. The presence of the flat response in the location $75.0mm \leq z \leq 85.0mm$, as shown in Fig. 3.4, corresponds the presence of an oscillation around 134 GHz in the RF spectrum plot. The appearance of this parasitic oscillation in the uptaper section of the geometry is considered to be due to the presence of the dynamic ACI phenomenon.

In addition to the peak at $\approx$134.72 GHz and at the desired oscillating frequency 140 GHz, other peaks may be treated as numerical noise as frequencies corresponding to these peaks (see Fig. 3.5) have not been found in the plots of equation 3.1 (see Fig. 3.4). As a consequence in order to confirm the existence of these unwanted spurious oscillations in the cavity output region, it is necessary to perform some more numerical investigations. [AIP+08] equipped the EURIDICE code with a broadband boundary condition and showed in corresponding single-mode calculations that ACI persists and it is not an artefact of the single-frequency boundary condition of the SELFT code. Of course, a realistic description of ACI phenomena requests for multi-mode simulations.

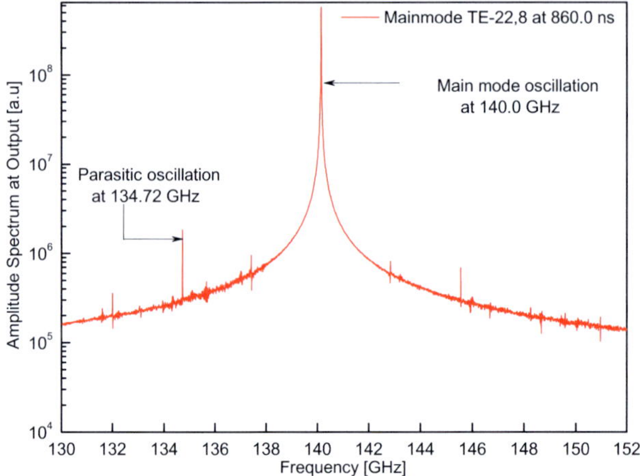

Figure 3.5: RF spectrum at the geometry output (see Fig. 3.2) at the end of the startup shown in Fig. 3.3. In this simulation result still only equation (2.29a) is solved for the particle motion while $u_{\parallel}$ has been treated as a constant value (single-mode calculation).

In the continuation of the above findings regarding the possibility of the appearance of dynamic ACI phenomena in the cavity output section, a few multi-mode simulations have also been performed using the unmodified version of the KIT SELFT code. In these calculations still only equation (2.29a) has been numerically solved. A similar approach (discussed above) has been adapted to plot the synchronization regions where the fulfilment of the interaction condition (3.1) exists along the cavity and uptaper geometry. In this case the chosen example was the 140 GHz $TE_{-28,8}$ mode W7-X gyrotron. In these multi-mode simulations three radial neighbour modes, which are relevant for our comparisons, have been taken into considerations. These modes may affect the excitation of the desired mode especially during the start-up. Moreover, all other modes will interact with each other through the electron beam during start-up but here we want to investigate the influence of radial-satellite modes as a ACI phenomena in the uptaper section of the geometry along with the excitation of the desire mode. Typical results describing regions of existence of the synchronization condition, obtained for the $TE_{-28,8}$ mode and its radial neighbour modes ($TE_{-28,7}$ and $TE_{-28,9}$) along the z-coordinate are shown in Fig. 3.6. It should be mentioned again that these simulated results are generated by using a uniform magnetic field in the electron motion as well as in the RF field differential equations and the non-uniform magnetic field only influences the detuning factor (as discussed earlier). The simulated electric field profiles (see Fig. 3.7) of the three modes $TE_{-28,8}$, $TE_{-28,7}$

and $TE_{-28,9}$ at an optimized operating point for the main operating mode $TE_{-28,8}$ clearly indicate that all three modes are exchanging power with the electron beam both in cavity and in the uptaper region. These simulated results have been treated as the initial estimation of the presence of parasitic/spurious oscillations in the cavity uptaper section. In order to have a clear understanding of the appearance of the oscillating frequencies which belong to either the main cavity operating mode or to other specific modes having different frequencies as spurious/parasitic oscillations which are suspected to be dynamic ACI, it is required to have an RF spectrum plot which is shown in Fig. 3.8.

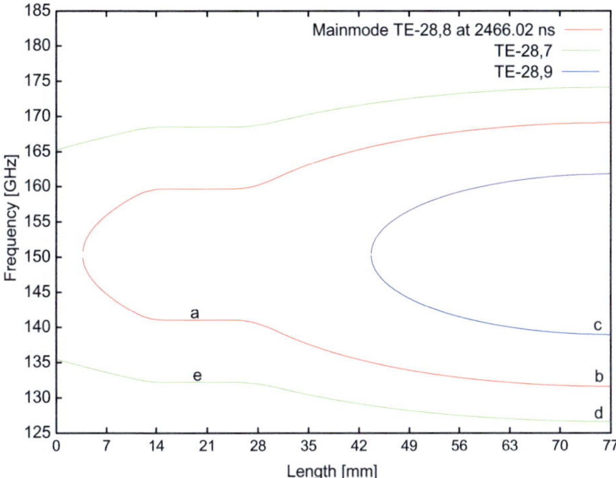

Figure 3.6: Estimated interaction frequencies for different modes, calculated from equations (3.1 and 3.7), for the parameters given in Fig. 3.7.

The plots of the electric field profile, which are shown in Fig. 3.7, depict the presence of weak amplitude modulations in the output section of the simulation domain, weaker than that of the single-mode simulation. These amplitude modulations in the electric field profiles clearly indicate the presence of undesired interactions between spent electron beam and RF travelling waves in the output section of the geometry. The RF spectrum of the simulated electric field profile is plotted in Fig. 3.8 in order to have a clear visualization of the appearance of the oscillating frequency of the main mode as well as any parasitic modes. From the RF spectrum plot it can be observed that in addition to the desired operating mode ($TE_{-28,8}$), denoted by ‘a’, other few modes also appear as parasitic/spurious modes having different oscillating frequencies which are denoted by ‘b’, ‘c’, ‘d’ and ‘e’ respectively. They can be suspected to be due to the presence of dynamic ACI phenomena. Out of all the above mentioned parasitic

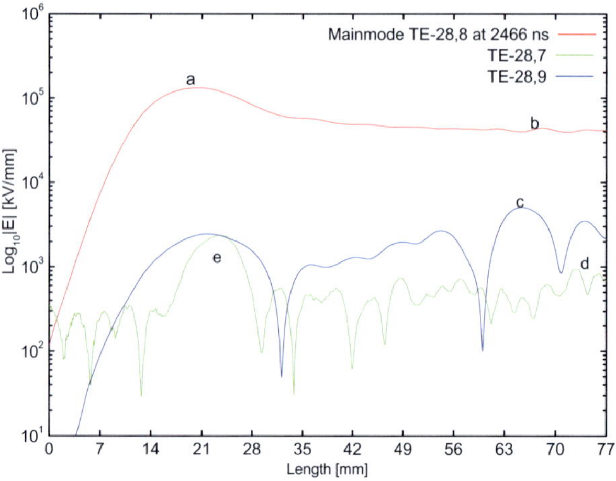

Figure 3.7: Calculated electric field profiles in a stationary operating point at $V_{beam} = 81.2$ kV, $I_{beam} = 39.0$ A, $\alpha = 1.3$, $B_0 = 5.56$ T and $R_{beam} = 10.3$ mm.

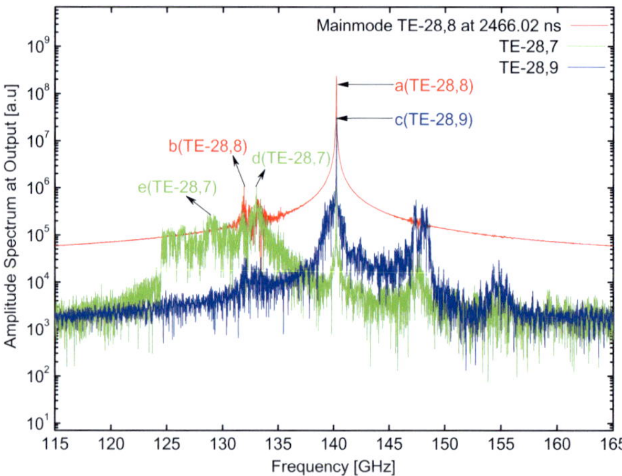

Figure 3.8: RF spectrum of the 3 relevant modes plotted at the end of the simulation domain (Fig. 3.7).

oscillations, present in the output section of the cavity geometry, few were found with a less significant magnitude in the RF spectrum plot whereas on the other hand the presence of parasitic oscillations in the uptaper section is clear from Fig. 3.6 and Fig. 3.7 respectively. As a consequence, parasitic oscillations with less significant amplitude may be due to numerical noise. Thus in order to confirm these oscillations as spurious oscillations in terms of the dynamic ACI phenomenon we have removed few approximations in the calculation using the unmodified version of the KIT SELFT code which will be described in the subsequent chapters of this thesis.

Looking at the RF spectrum plot of the field profiles (see Fig. 3.8), several beam-wave interactions have been identified which are labeled as 'b', 'c', 'd' and 'e'. These beam-wave interactions points also correspond to the presence of the flat responses in Fig. 3.6 as well as modulated electric field profiles (see Fig. 3.7). Some of the salient points have been drawn from the RF spectrum plots (see Fig. 3.8) which are as follows:

a) Main line of the $TE_{-28,8}$ mode at 140.2 GHz: This is the desired cavity beam-wave interaction corresponding to the straight portion of the resonator.

b) Spurious line of the $TE_{-28,8}$ mode at 131.4 GHz: This interaction corresponds to a broad region in the uptaper ($65.0mm \leq z \leq 77.0mm$) with relatively constant interaction frequency. This interaction may be labeled as dynamic ACI phenomenon. The intensity of this interacting mode is much less in comparison with the main mode intensity and so it is necessary to have some more investigations on the appearance of the dynamic ACI by considering more realistic cases like a non-uniform distribution of the magnetic field profile in the simulation domain particularly in the uptaper section of the cavity.

c) Main line of the $TE_{-28,9}$ mode at 140.2 GHz which is similar to b), but for a different mode: The frequency seems to be locked to the main mode's oscillation.

d) Weak and noisy line for the $TE_{-28,7}$ mode at 124 GHz which is again similar to b), but weakly coupled to the electron beam: This interaction line corresponds to the flat response in the location $65.0mm \leq z \leq 77.0mm$ as shown in Fig. 3.6.

e) An interaction line of the $TE_{-28,7}$ mode at 134.7 GHz corresponds to the main cavity interaction region but with a weak coupling to the electron beam.

Further refinements in the simulations have been done by considering the influence of the cavity beam-wave interaction after the straight portion of the cavity (i.e. in the uptaper section) on the electron beam parameters by calculating the pitch factor (α) as well as γ for each particle as a function of the velocity components ($u_\perp$ and $u_\parallel$). These velocity components were in turn calculated by considering a non-uniform distribution of the magnetic field profile in the time-dependent differential equations required for the self-consistent beam-wave interaction calculations (see equations (2.29) and (2.8)) (for more details see chapter 4).

In all the above simulation results it has been considered that each macro-particle has the same longitudinal velocity $u_{\parallel}$ but this is not true when the magnetic field is not considered to have a constant value. The consideration of the influence of the change of the magnetic field in the beam-wave interaction calculations is necessary to simulate the interaction phenomenon, particularly near the uptaper section of the geometry, as close to the reality as possible and also in order to correlate the simulated results with the experimental one obtained with the reference to the study of the dynamic ACI performed at KIT [KAC$^+$10]. As a consequence, it is required to consider that each particle has its own γ and velocity components $u_{\perp}$ and $u_{\parallel}$ depending upon the non-uniform distribution of the magnetic field. As a result each of them has an individual curve of resonance frequencies by means of which the regions of synchronism between RF wave and electron beam can be depicted. The result of possible interaction frequencies for each electron as well as for each mode are shown in Fig. 3.9 plotted along the z-coordinate of the cavity geometry (140 GHz $TE_{-28,8}$ mode W7-X gyrotron having the following beam parameters: $V_{beam} = 81.8$ kV, $I_{beam} = 43.2$ A, $\alpha = 1.18$, $B_0 = 5.615$ T and $R_{beam} = 10.17$ mm). In these calculations $\beta_{\parallel}$ has been obtained by using the adiabatic approximation (equation (3.5), see section 3.2) for each macro-particle calculated at each z value using the analytical formula equation (3.5) in which the values of $\beta_{\perp}$ for each macro-particle are obtained from the interaction calculations by solving only equation (2.29a). These locally varying $\beta_{\parallel}$ were used in the solutions of equation (3.1) in order to determine the regions where the synchronization condition may exist along the whole simulation domain.

The solutions shown in Fig. 3.9 correspond to the two crossing points of each mode's hyperbola in the Brillouin-Diagram (see chapter 1 and for more details see [Ker96]) with the electron beam line having different γ (obtained from the interaction calculations) in the calculations of Ω_c for different particles. Fig. 3.9 has been plotted while considering 900 micro-particles and 13 $TE_{m,p}$ modes with different directions of rotation (see Fig. 3.11). It has already been mentioned that the longer straight regions with nearly constant frequency vs. z can be considered for regions suitable for a net energy exchange between RF wave and electron beam. Thus from Fig. 3.9, a frequency of 140 GHz for the main mode $TE_{-28,8}$ has been estimated from the lower section of the curves in the cylindrical cavity region ($18.0mm \leq z \leq 22.0mm$), as desired. But on the other hand, almost flat frequency responses around 132 GHz are also found in the uptaper region ($65.0mm \leq z \leq 77.0mm$) with different $TE_{m,p}$ modes which may be considered as undesired oscillations suspected to be due to the dynamic ACI phenomenon. Thus it can be assumed that this undesired additional interaction region is the reason for the dynamic ACI phenomenon in these simulations which require a deeper investigation.

The upper high-frequency gyro-TWT branches in Fig. 3.9 are broadened due to strong sensitivity to different $u_{\parallel}$ (Doppler broadening). Thus they are much less prone to dynamic ACI.

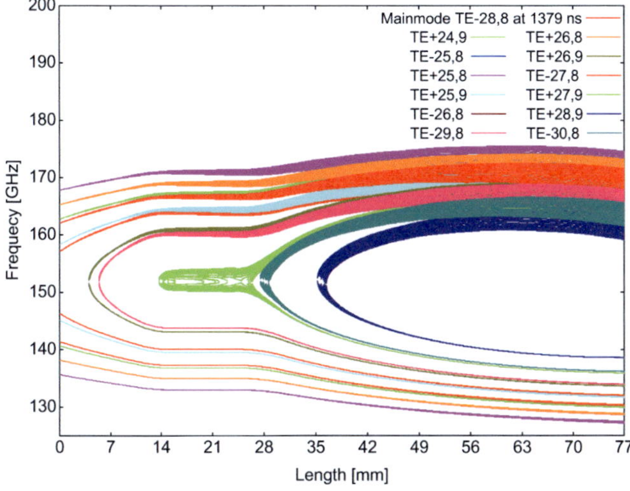

Figure 3.9: Estimated interacting frequencies curves for 900 macro-particle and 13 $TE_{m,p}$, modes, having different γ and $\beta_{\parallel}(z)$ for each particle, at parameters shown in Fig. 3.10.

In the next numerical experiment, the values of $\beta_{\parallel}$, calculated using the adiabatic constant which depends on the non-uniform distribution of the magnetic field profile, have been used in the time-dependent self-consistent multi-mode simulation using the unmodified version of the SELFT code. In this numerical experiment again only equation (2.29a) has been solved numerically. Electric field plots have been generated and are shown in Fig. 3.10. The presence of undesired oscillations in the output section of the simulation domain, as depicted in Fig. 3.10, have been observed due to the presence of parasitic modes $TE_{-27,8}$, $TE_{+26,9}$ and $TE_{+26,8}$. This indicates power exchange between traveling RF wave and the spent electron beam in the uptaper section of the cavity geometry (see Fig. 3.10). The electric field strength of the other competing modes has been observed significantly lower ($\approx$less than 10^3) than the strength of the main operating mode electric field profile and hence may be considered at the noise level. Thus other competing modes do not contribute in the present investigations. This parasitic mode $TE_{-27,8}$ clearly features the dynamic ACI phenomenon in the uptaper section and indicates that the consideration of a non-uniform magnetic field distribution in the differential equations is necessary for the realistic investigations of the dynamic ACI phenomenon in the cavity output section.

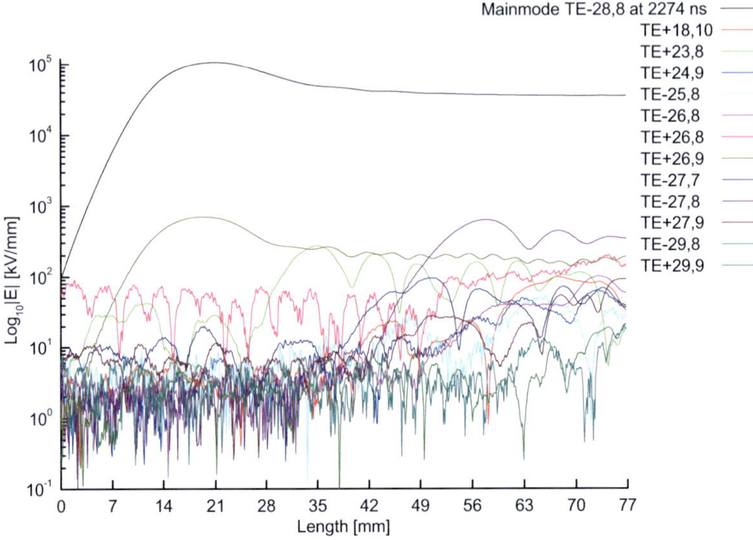

Figure 3.10: Calculated electric field profiles in a stationary operating point at $V_{beam} = 81.8$ kV, $I_{beam} = 43.2$ A, $\alpha = 1.18$, $B_0 = 5.615$ T and $R_{beam} = 10.17$ mm . $\beta_{\shortparallel}(z)$ are used in calculating electric field profiles.

The RF spectrum plot at the end of the simulation domain has been calculated at the simulation time $t_s = 2200$ ns and is shown in Fig. 3.11. From this plot one can confirm that in addition to the desired ≈140 GHz oscillation of the cavity, other parasitic oscillations like at ≈134.56 GHz ($TE_{+26,9}$ mode), ≈130.14 GHz ($TE_{-27,8}$ mode) and at ≈128.19 GHz ($TE_{+26,8}$) are also observed but with a much lower magnitude as compared to the main mode oscillation. These undesired parasitic frequencies also correspond to flat responses of the resonance frequency curves (see Fig. 3.9). The consideration of the adiabatic approximation in the calculation of the electron parallel velocity which includes the influence of the non-uniform distribution of the magnetic field profile along the simulation domain provides a better estimation of parasitic/spurious oscillations in the uptaper section. This observation was evident from the RF spectrum plot (see Fig. 3.11) where the magnitude of the intensity of the spurious oscillations, in the uptaper section of the geometry, was more prominent than in the RF spectrum plots (see Fig. 3.8), which were obtained from the previously discussed numerical experiment. This confirms that the inclusion of the influence of the non-uniform distribution of the magnetic field directly in the time-dependent gyrotron beam-wave interaction differential equations (i.e. equations

(2.8) and (2.29)) is necessary in order to make proper investigations in terms of the appearance of the dynamic ACI particularly in the uptaper section where the non-uniform distribution of the magnetic field is more significant. Further numerical investigations have been performed under the consideration of the non-uniform magnetic field distribution directly in the solutions of the differential equations which are responsible for the gyrotron beam-wave interaction calculations in the SELFT code (see chapter 6).

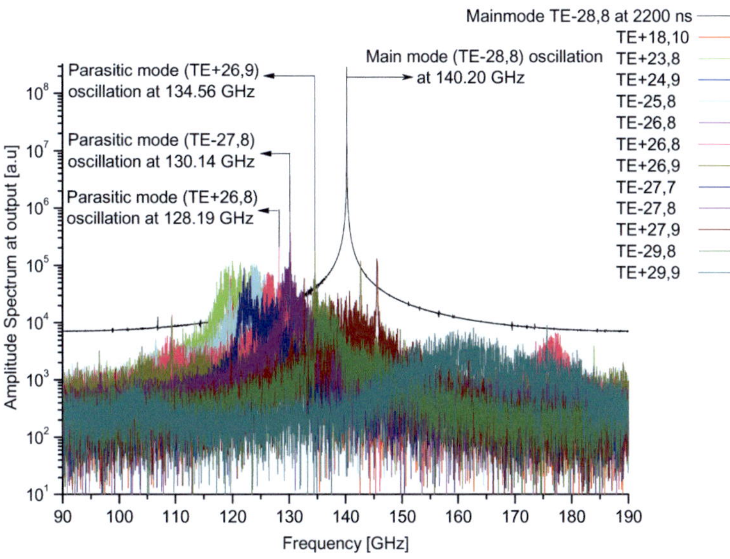

Figure 3.11: RF spectrum at the end of the simulation domain of all the simulated modes. Frequencies of the main mode and parasitic modes correspond to Fig. 3.10.

4 Numerical implementation and proposed modifications in the SELFT code

We now discuss the numerical implementation of the described differential equations applied in the SELFT code as well as the modifications made in the required differential equations with reference to make more realistic investigations on the dynamic ACI phenomena in the uptaper section of gyrotron cavities under the consideration of a non-uniform distribution of the magnetic field. In the SELFT code, finite difference methods with several approximations are used in order to make simulations fast and accurate. These implemented approximations are described in [Ker96]. Particularly the field profile equation (2.8) and electron motion equation (2.29) are solved by using the Crank-Nicholson method as well as Predictor-Corrector method, respectively, and are described in the later part of this chapter.

In this chapter, a discretized form of the differential equations (see sections 4.2 and 4.3), responsible for the actual beam-wave interaction calculations, and their implementations in the existing SELFT code, which includes the influence of the non-uniform distribution of the magnetic field, are discussed for numerical solution of the physical behaviour of the beam-wave interaction process. In this course, the stationary cold cavity field profile along with the implemented boundary conditions in the simulations of the RF field profile, stationary as well as non-stationary, are also discussed followed by discussions on the numerical aspects of the electron momentum equation. Finally, a modification with respect to the reference frequency is also been discussed in which a slowly varying reference frequency has been used for the calculations of the slow-phase-variable. The axial z-coordinate as well as the time coordinate are discretized by z_n and t_i respectively which are given by:

$$z_n = z_{in} + n \cdot \Delta z, \quad t_i = t_{i-1} + \Delta t \quad n = 0, 1, 2, \ldots \tag{4.1}$$

Here Δz and Δt denote the step size of the axial as well as time coordinate discretization and the corresponding sample of the field profile can be given as follows:

$$\vec{f}_{k,n,i} = \vec{f}_k(z_n, t_i) \tag{4.2}$$

The index k is the same as described in chapter 2 but in general it is ignored for the sake of simplicity.

4.1 Implemented numerical method for the stationary cold cavity field profile

In the SELFT code, non-stationary multi-mode simulations have been carried out by considering the stationary cold cavity field profile along the resonator axis as an initial RF field profile which is obtained by considering the excitation term in the field profile equation equal to zero i.e. $\vec{R}_k = 0$ (see equation 2.10). The cold cavity field profile has already been proposed in [FR81]. In the SELFT code these cold cavity field profiles are effectively solved by using the Numerov-Algorithm [Sch68]. The angular frequency ω, which has been included in the calculation of $k_{\shortparallel}$ for each $\mathrm{TE}_{m,p}$ mode, has been expressed as a complex quantity $\vec{\omega}$ in the SELFT code [Ker96], wherein the imaginary part describes the rate of the decay process and is given by:

$$\vec{\omega} = \omega\left(1 + j\frac{1}{2\cdot Q_{dif}}\right) \tag{4.3}$$

Using (4.3), the equation for the cold cavity field profile (2.10) can be expressed as follows:

$$\frac{\partial^2}{\partial z^2}(\vec{f}(z)) = -\vec{k}_{\shortparallel}^2 \cdot \vec{f}(z) = \left(-\frac{\omega^2}{c^2} + \frac{\omega^2}{4c^2 Q_{dif}^2} - k_{\perp}^2(z) - j\frac{\omega^2}{c^2 Q_{dif}}\right)\vec{f}(z) \tag{4.4}$$

The quantity $k_{\perp}^2(z)$ in equation (4.4) can be obtained, at each discretized value of z, by using the relation $k_{\perp}(z) = \chi_{m,p}/R_0(z)$ and as a result the above cold cavity field profile equation can be discretized numerically in equidistance along the z-axis by using the Numerov-Algorithm [Sch68] which is given by:

$$\begin{aligned}\frac{\vec{f}_{n+1} - 2\vec{f}_n + \vec{f}_{n-1}}{\Delta z^2} &= \frac{1}{12}\left(\vec{k}_{\shortparallel,n+1}^2 \vec{f}_{n+1} + 10\vec{k}_{\shortparallel,n}^2 \vec{f}_n + \vec{k}_{\shortparallel,n-1}^2 \vec{f}_{n-1}\right)\\ \vec{f}_{n+1} = 2\vec{f}_n - \vec{f}_{n-1} &- \frac{\Delta z^2}{12}\left(\vec{k}_{\shortparallel,n+1}^2 \vec{f}_{n+1} + 10\vec{k}_{\shortparallel,n}^2 \vec{f}_n + \vec{k}_{\shortparallel,n-1}^2 \vec{f}_{n-1}\right)\end{aligned} \tag{4.5}$$

The above equation (4.5) has been solved and stored at each discretized value of the z-axis which is further used as the initial values for the non-stationary multi-mode self-consistent simulation. For more details see [Ker96, and references therein]. The solution of equation (4.5) has been associated with the boundary condition that is applied in the SELFT code which can be defined as follows:

$$\begin{aligned}\frac{\partial}{\partial z}\left(\vec{f}(z_{in})\right) &= +j\frac{1-\vec{\Gamma}_{in}}{1+\vec{\Gamma}_{in}}\vec{k}_{\shortparallel}\vec{f}(z_{in})\\ \frac{\partial}{\partial z}\left(\vec{f}(z_{out})\right) &= -j\frac{1-\vec{\Gamma}_{out}}{1+\vec{\Gamma}_{out}}\vec{k}_{\shortparallel}\vec{f}(z_{out})\end{aligned} \tag{4.6}$$

where $\vec{\Gamma}_{in}$ and $\vec{\Gamma}_{out}$ denotes the reflection coefficient at the input and output of the simulation domain respectively. The discretized form of the above boundary condition equation (4.6) has been used in the SELFT code for the solution of equation (4.5). The difference forms of equation (4.6), obtained only at the input and output of the simulation domain have been described as follows:

$$\frac{\vec{f}_1-\vec{f}_{-1}}{2\Delta z}=+j\frac{1-\vec{\Gamma}_{in}}{1+\vec{\Gamma}_{in}}\vec{k}_{\parallel,0}\vec{f}_0 \tag{4.7a}$$

$$\frac{\vec{f}_{n_{out}+1}-\vec{f}_{n_{out}-1}}{2\Delta z}=-j\frac{1-\vec{\Gamma}_{out}}{1+\vec{\Gamma}_{out}}\vec{k}_{\parallel,n_{out}}\vec{f}_{n_{out}} \tag{4.7b}$$

Since in the gyrotron simulation, the $\mathrm{TE}_{m,p}$ mode should be at the cavity resonant frequency placed at the input of the resonator or somewhere in between the downtaper and straight portion of the cavity (see Fig. 1.7) which is well below the cut-off frequency so, in view of the above, the calculation of the boundary condition at the input of the simulation domain is not taken into account in this chapter and for more details see [Ker96]. In the SELFT code the solution for the cold cavity field profile has been obtained by integrating equation (4.5) over the axial coordinate z while using the boundary condition (4.7b) optimized for the particular given resonant angular frequency (ω) and quality factor (Q_{dif}) in such a way that the reflection coefficient at the output of the simulation domain must satisfy the following relation, which is given by:

$$\vec{\Gamma}_{out}\left(\omega,Q_{dif}\right)=\frac{2\Delta z\vec{k}_{\parallel,n_{out}}\vec{f}_{n_{out}}-j\left(\vec{f}_{n_{out}+1}-\vec{f}_{n_{out}-1}\right)}{2\Delta z\vec{k}_{\parallel,n_{out}}\vec{f}_{n_{out}}+j\left(\vec{f}_{n_{out}+1}-\vec{f}_{n_{out}-1}\right)} \tag{4.8}$$

Generally the value of the output reflection coefficient ($\vec{\Gamma}_{out}\left(\omega,Q_{dif}\right)$)at the desired frequency of oscillation is required to be optimized in order to get the require beam-wave interaction simulation using the existing SELFT code but ideally it should be considered equal to zero i.e.

$$\vec{\Gamma}_{out}\left(\omega,Q_{dif}\right)=0 \tag{4.9}$$

Optimization has been performed using a non-linear optimization techniques [PTV+92] with different values of quality factor in order to obtain a minimum absolute value of the reflection factor $\mid\vec{\Gamma}_{out}\mid$ or less than the predefined limit (for example 0.01) for the calculation of a single-mode cold cavity field profile. It should be mentioned that for the same transverse mode indices k (k stands for m and p) several solutions may occur which correspond to different axial mode indices n [Ker96]. Thus one can perform optimization for getting the required value of n by considering the minimum value of the frequency, for example the cut-off frequency of the resonator. It was also observed that the resonant frequency increases with increasing n. In the SELFT simulation n is generally considered to be 1 (first axial mode).

4.2 Numerical methods implemented for stationary and non-stationary beam-wave interaction simulation

A detailed description of the numerical methods implemented in the unmodified version of the SELFT code for stationary as well as non-stationary simulations, with constant magnetic field, has been discussed in [Ker96]; a brief discussion is presented in this section for a better reference. In many publications for example in [BJ87, BJ92] it has been reported that self-consistency in the simulation is necessary for the most reliable beam-wave interaction studies with or without non-uniform distribution of the magnetic field. The self-consistent approach is certainly provided by solving both equations, i.e. the RF field equation equation (2.8) and electron motion equation (2.29), simultaneously in such a way that the changing beam parameters (like electron transverse as well as longitudinal velocities) have an influence on the evolution of the RF field profile through the calculation of the excitation term ($\vec{R}_n$) which is directly influenced by the beam current ($I_{B,j}$), relativistic factor (γ_j) as well as electron transverse and longitudinal velocities.

In the case of stationary self-consistent single-mode simulation, the stationary RF field profile equation (2.10) has been discretized by using a 'Leapfrog' scheme [PTV+92] i.e.

$$
\begin{aligned}
&\frac{\partial^2}{\partial z^2}\left(\vec{f}(z)\right)+\vec{k}_{\shortparallel}^2\cdot\vec{f}(z)=\vec{R}_n \\
&\frac{\partial}{\partial z}\vec{f}_{n+\frac{1}{2}}=\frac{\partial}{\partial z}\vec{f}_{n-\frac{1}{2}}+\Delta z\left(\vec{R}_n-\vec{k}_{\shortparallel,n}^2\cdot\vec{f}_n(z)\right),\quad \vec{f}_{n+1}=\vec{f}_n+\Delta Z\frac{\partial}{\partial z}\vec{f}_{n+\frac{1}{2}}
\end{aligned}
\tag{4.10}
$$

In order to have a self-consistent stationary single-mode calculation, the discritization of the electron momentum equation has also been carried out and is described in section 4.3 whose solutions have a direct influence on the discretized stationary field equation (4.10) through the excitation term $\vec{R}_n$. It should also be mentioned that in the case of a stationary solution of a single-mode calculation i.e. in the solution of the equation (4.10), the same boundary condition has been implemented as was used for the evolution of the cold field profile (see equation 4.5) (discussed in section 4.1). The simulation results obtained with the self-consistent stationary single-mode calculations provide a rough estimation of the output power and electronic as well as RF efficiencies for fixed values of the beam parameters. Then these estimated values can be used as input for self-consistent non-stationary multi-mode calculations. As in the case of self-consistent calculations, for given parameters, several solutions can be possible and thus this property can also be used for parametric studies. For more details see [Ker96] and references therein. Moreover all the above implementations have been treated as input for the non-stationary self-consistent multi-mode simulations.

In the time-dependent self-consistent beam-wave interaction calculations, the computational time, due to the inclusion of multi-mode features influenced by the non-uniform distribution of the magnetic field, increases by 40% as compared to the calculations reported in [Ker96] where the magnetic field is taken as a constant value and where a typical computational time is $\approx$20 hours. This is because most of the computing time is consumed for solving the particle motion equation in which particle momenta (transverse as well as longitudinal) are also influenced by the non-uniform distribution of the magnetic field as compared to [Ker96]. Multi-mode calculations are therefore exclusively performed non-stationary and self-consistently under the influence of the non-uniform distribution of the magnetic field.

In the existing SELFT code, due to the consideration of multi-mode features, the frequency ω_k is no longer the oscillation frequency as compared to the stationary self-consistent single-mode simulation but it has been estimated by a fixed averaged value rate and should preferably be close to the resonant frequency [Ker96]. The deviation of the oscillation frequency from the averaged frequency is described by the difference between the timings of the phases of the associated field profiles present at time steps i and $i-1$ respectively divided by the time step width and has been determined by [Ker96]:

$$\omega = \omega_k + \frac{\angle\left(\vec{f}_{k,i}\right) - \angle\left(\vec{f}_{k,i-1}\right)}{\Delta t} \tag{4.11}$$

where $\angle$ represents the phase angle of the field profiles observed at time i and $i-1$ respectively, and Δt represents the discritization time step size. In the time-dependent multi-mode simulations using the unmodified SELFT code, the oscillating frequency obtained by using equation (4.11) provides a sufficiently well defined frequency and it has also been observed that it has very less influence on the estimation of power and efficiencies.

In order to perform time-dependent self-consistent multi-mode simulations under the influence of a non-uniform distribution of the magnetic field, it is of course necessary also to solve the coupled differential equations of the field profile (2.8) and the electron momentum equation (2.29) simultaneously which includes the term containing the effects of the non-uniform distribution of the magnetic field. The difference form of the electron momentum equation is described in the next section 4.3. For the numerical implementation of the equation of the RF field profiles in the modified version of the SELFT code, an implicit solver is used because due to the Courant condition [PTV+92] there is no longer a limitation on the time step size. Due to the implicit solver, the numerical stability region is much larger than for an explicit one. In this case the maximum time step is then determined, not due to the numerical instabilities, but by the time behaviour of the physical quantities. In the modified version of the SELFT code a partially implicit Crank-Nicholson scheme [PTV+92] has been implemented for solving the RF field profile equation (2.8) [Ker96]. For better understanding a brief description of the numerical aspects of the Crank-Nicholson scheme is introduced below.

Numerical aspects of Crank-Nicholson scheme

The Crank-Nicholson method is a finite difference method used for numerically solving diffusion type partial differential equations. It has second-order accuracy both in space as well as in time because it combines the accuracy of both implicit and explicit FTCS (Forward Time Centered Space) schemes. Stability criteria for this numerical method can be seen from Von Neumann stability analysis. It is a standard textbook result that this scheme is unconditional stable, see [PTV+92] for details. However, the approximate solutions can still contain (decaying) spurious oscillations if the ratio of time step Δt to the square of space step Δz^2 is large. For this reason, whenever large time steps or high spatial resolution is necessary, the less accurate backward Euler method is often used, which is both stable and immune to oscillations. The Crank-Nicholson method is based on central difference in space domain and the trapezoidal rule in time domain, giving second-order convergence in time. Let us consider a one-dimensional diffusion equation, as an example, for illustration purpose.

$$\frac{\partial f}{\partial t} = D\frac{\partial^2 f}{\partial z^2}$$

Applying the Crank-Nicholson scheme for the discretization of the above diffusion equation, we can obtain the value of the function f at each discrete point of the geometry.
With $\alpha = \frac{D\Delta t}{2\Delta Z^2}$ and rewriting the above equation we get:

$$-\alpha f_{n+1,i+1} + (1+2\alpha) f_{n,i+1} - \alpha f_{n-1,i+1} = \alpha f_{n+1,i} + (1+2\alpha) f_{n,i} - \alpha f_{n-1,i}$$

The above equation is a tridiagonal problem in which $f_{n,i+1}$ can be efficiently solved by using the tridiagonal matrix algorithm.
Similarly the RF field profile equation (2.8) has been discretized by using the above explained Crank-Nicholson algorithm and has been used in the modified SELFT code in order to have RF field values at each point of the resonator along the axial coordinate; the discretized form of equation (2.8) can be described by [Ker96]:

$$\begin{aligned} &\frac{\vec{f}_{k,n+1,i+1} - 2\vec{f}_{k,n,i+1} + \vec{f}_{k,n-1,i+1}}{2\Delta z^2} + \frac{\vec{f}_{k,n+1,i} - 2\vec{f}_{k,n,i} + \vec{f}_{k,n-1,i}}{2\Delta z^2} \\ &+ \underbrace{\left[\frac{\omega_k^2 - \omega_{\perp,k}^2}{c^2}\right]_n}_{A_{k,n}} \vec{f}_{k,n,i} - j\underbrace{\frac{2\omega_k}{c^2}}_{B_k} \frac{\vec{f}_{k,n,i+1} - \vec{f}_{k,n,i}}{\Delta t} = \vec{R}_{k,n,i} \end{aligned} \tag{4.12}$$

The above equation (4.12) has been solved and implemented in the modified version of the SELFT code by using the boundary condition given by equation (4.7). The equation (4.12) has been recasted into the implicit matrices form and can be described by:

$$\left(\vec{\mathrm{M}}_{k,i} - j\frac{B_k}{\Delta t}\mathbf{I}\right) \cdot \vec{\mathrm{f}}_{k,i+1} = \vec{\mathrm{R}}_{k,i} - \left(\vec{\mathrm{M}}_{k,i} + j\frac{B_k}{\Delta t}\mathbf{I} + \mathbf{A}_k\right) \cdot \vec{\mathrm{f}}_{k,i} \tag{4.13}$$

The matrices, vectors and constants used in equation (4.13) are described in equation (4.14).

$$\vec{\mathrm{f}}_{k,i} = \begin{pmatrix} \vec{f}_{k,n_{out},i} \\ \vdots \\ \vec{f}_{k,n,i} \\ \vdots \\ \vec{f}_{k,0,i} \end{pmatrix} \mathbf{A}_k = \begin{Bmatrix} A_{k,n_{out}} & & & & 0 \\ & \ddots & & & \\ & & A_{k,n} & & \\ & & & \ddots & \\ 0 & & & & A_{k,0} \end{Bmatrix} \vec{\mathbf{R}}_k = \begin{Bmatrix} \vec{R}_{k,n_{out}} & & & & 0 \\ & \ddots & & & \\ & & \vec{R}_{k,n} & & \\ & & & \ddots & \\ 0 & & & & \vec{R}_{k,0} \end{Bmatrix}$$

$$\vec{\mathbf{M}}_k = \frac{1}{\Delta, z^2} \begin{Bmatrix} -1-\vec{C}_{out,i} & 1 & 0 & \\ \ddots & \ddots & \ddots & \\ \frac{1}{2} & -1 & \frac{1}{2} & \\ & \ddots & \ddots & \ddots \\ 0 & 1 & -1-\vec{C}_{in,i} & \end{Bmatrix}$$

$$A_{k,n} = \left[\frac{\omega_k^2 - \omega_{\perp,k}^2}{c^2}\right]_n \quad B_k = \frac{2\omega_k}{c^2} \quad \vec{C}_{in,out,i} = j\vec{k}_{||,k,i} \frac{1-\vec{\Gamma}_{in,out}}{1+\vec{\Gamma}_{in,out}} \Delta z \tag{4.14}$$

In the modified SELFT code, the above matrix equations are solved at each time step after having done a complete calculation of the RF field excitation term $\vec{R}_{k,n,i}$ at every point of the resonator. In contrast to the calculations reported in [Ker96] in the calculation of the above matrix equation (4.13), the $\vec{R}_{k,n,i}$ contain terms which are directly influenced by the non-uniform distribution of the magnetic field profile. In the modified SELFT code, the inversion of the tridiagonal matrix has been performed by a standard subroutine called TRIDAG written in the complex form [PTV+92, Ker96]. In order to make investigations regarding the occurrence of dynamic ACI phenomena in the uptaper section of the geometry, for which the change of the magnetic field is more significant, the quantity D_n, as shown in equation (4.15) where Mag_n (magnetic field value at the nth discrete point) has been included in the calculation of the excitation term of the RF field profile in contrast to the same quantity reported in [Ker96]. As a consequence the above matrix equation (4.13) has been modified accordingly. This is one of the important modifications done in the modified SELFT code towards the investigations of dynamic ACI and is one of the major differences between the unmodified and modified version of the SELFT code.

$$\vec{R}_{k,n,i} = \mu_0 \sum_j \frac{I_{B,j}}{u_{||,j,n,i}} \left[G_{e,k,j,n,i} e^{-jm\varphi_{e,i}} \cdot e^{j(\omega_k - \Omega_f)t}\right]_n \cdot \left(j\Omega_0 \cdot \frac{\vec{p}_{j,n,i}}{\gamma_{j,n,i}} + \underbrace{\left(\frac{\gamma_{j,n,i}^2 - 1}{u_{||,j,n,i}^2}\right) \overbrace{\left(\frac{1}{2B_{R,z,n}} \frac{dB_{R,z,n}}{dz}\right)}^{Mag_n} \cdot \frac{\vec{p}_{j,n,i}}{\gamma_{j,n,i}}}_{D_n} \right) \tag{4.15}$$

In order to implement the effect of the non-uniform distribution of the magnetic field in the modified SELFT code, the magnetic field profile values have been obtained from the external code ESRAY existing at KIT [Ill97]. The quantity Mag_n in equation (4.15) has been equidistantly discretized along the axial coordinate, having a second order accuracy by using equation (4.16). Finally discretized values have been used in the calculations of $\vec{R}_{k,n,i}$ at each point of the resonator.

$$\frac{1}{2B_{R,z}}\frac{dB_{R,z}}{dz} = \frac{1}{B_{R,n}}\left\{\begin{array}{lr} (-1.5B_{R,1}+2B_{R,2}-0.5B_{R,3})/\Delta z & \text{when} \quad n=1 \\ (B_{R,n+1}-B_{R,n-1})/2\Delta z & \text{when} \quad 1<n<nz-1 \\ (1.5B_{R,nz}-2B_{R,nz-1}+0.5B_{R,nz-2})/\Delta z & \text{when} \quad n=nz \\ nz = \text{Maximum number of discrete points} & \end{array}\right\} \tag{4.16}$$

In the existing SELFT code finally the amplitude of the field profiles has been used in order to determine the output power ($p_{dif,k}$) according to the equation determined by equation (4.17). The detailed derivation of this equation can be found in [Ker96]:

$$p_{dif,k} = \left(\frac{1-\left|\vec{\Gamma}_{in,out,k}\right|^2}{1+2\mathrm{Re}\left|\vec{\Gamma}_{in,out,k}\right|+\left|\vec{\Gamma}_{in,out,k}\right|^2}\right)\cdot\left(\frac{k_{\shortparallel,k}}{2\mu_0\omega}\left|\vec{f}_k(z_{in,out})\right|^2\right) \tag{4.17}$$

4.3 Numerical methods implemented for solution of the Lorentz force differential equation

The Lorentz force equation, both in transverse as well as longitudinal direction (2.29), has been resolved in the modified SELFT code in parallel with the RF field profile obtained from equation (4.13) at each time step in order to obtain a self-consistent solution. In equation (2.29), the calculated RF field values are included in the acceleration term $\vec{a}(r_j,\varphi_j)$ (2.41a). The discretized form of the electron motion momentum equation has been accomplished by using a Heun's Predictor and Corrector method [PTV+92, EMR93]. For better understanding a brief introduction of the Predictor and Corrector method is introduced below.

Numerical aspects of the Predictor-Corrector method

A Predictor-Corrector method is a numerical method that utilizes two steps to determine the correct approximated solution. First, the prediction step, in which Euler's Method (explicit method) is used. It does a rough approximation of the desired quantity; in this case the electron momentum values are approximated. Secondly, the corrector step refines the initially approximated values using the Trapezoidal Method (implicit method) to obtain a better approximation to the true solution. A differential equation as an example has been considered for illustration purpose. Let us consider a first order partial differential equation:

$$p\prime(z) = f(z,p(z)), \quad p(z_n) = p_n$$

Then by using the Heun's Predictor-Corrector method, the above equation can be discretized as:

$$\tilde{p}_{n+1} = p_n + \Delta z \cdot f(z_n, p_n) \quad \text{(Euler's method-Intermidiate value)}$$

$$p_{n+1} = p_n + \frac{\Delta z}{2}\left(f(z_n, p_n) + f(z_{n+1}, \tilde{p}_{n+1})\right)$$

$$\text{(Trapezoidal method-Final approximated value at next } \Delta z)$$

The predictor and the corrector steps have local truncation errors of O(Δz^2) and O(Δz^3), respectively. The efficiency of the iterative corrector depends on the accuracy of the initially predicted value.

As already mentioned above, the electron motion momentum equation, both in transverse as well as longitudinal direction, (2.29) has been discretized by using the above explained Heun's Predictor and Corrector method and implemented in the modified SELFT code. The discretized version of equation (2.29) has been described by equation (4.18), which is given by:

Predictor Step

$$\left.\begin{aligned} \tilde{\vec{p}}_{n+1} &= \vec{p}_n + \Delta z \left[\left\{ \frac{\vec{a}_n \gamma_n + j\left(\Omega_0 - \Omega_f \gamma_n\right)\vec{p}_n}{u_{\shortparallel,n}} \right\} + j \cdot \mathrm{Mag}_n \cdot \vec{p}_n \right] \\ \tilde{u}_{\shortparallel,n+1} &= u_{\shortparallel,n} + \Delta z \left\{ \left(-\frac{|\vec{p}_n|^2}{u_{\shortparallel,n}} \right) \cdot \mathrm{Mag}_n \cdot \vec{p}_n \right\} \end{aligned}\right\}$$

$$\tilde{\gamma}_{n+1} = \frac{1}{\sqrt{1-(v/c)^2}} = \sqrt{1 + \tilde{u}^2_{\shortparallel,n+1} + |\tilde{\vec{p}}_{n+1}|^2}$$

Corrector Step

$$\left.\begin{aligned} \vec{p}_{n+1} &= \vec{p}_n + \frac{\Delta z}{2} \left[\begin{aligned} &\left(\left\{ \frac{\vec{a}_{n+1}\tilde{\gamma}_{n+1} + j\left(\Omega_0 - \Omega_f \tilde{\gamma}_{n+1}\right)\tilde{\vec{p}}_{n+1}}{\tilde{u}_{\shortparallel,n+1}} \right\} + j \cdot \mathrm{Mag}_{n+1} \cdot \tilde{\vec{p}}_{n+1} \right) \\ &\quad + \underbrace{\left(\left\{ \frac{\vec{a}_n \gamma_n + j\left(\Omega_0 - \Omega_f \gamma_n\right)\vec{p}_n}{u_{\shortparallel,n}} \right\} + j \cdot \mathrm{Mag}_n \cdot \vec{p}_n \right)}_{=\left(\tilde{\vec{p}}_{n+1} - \vec{p}_n\right)/\Delta z} \end{aligned} \right] \\ u_{\shortparallel,n+1} &= u_{\shortparallel,n} + \frac{\Delta z}{2} \left[\left\{ -\frac{|\tilde{\vec{p}}_{n+1}|^2 \cdot \mathrm{Mag}_{n+1} \cdot \tilde{\vec{p}}_{n+1}}{\tilde{u}_{\shortparallel,n+1}} \right\} + \underbrace{\left\{ -\frac{|\vec{p}_n|^2 \cdot \mathrm{Mag}_n \cdot \vec{p}_n}{u_{\shortparallel,n}} \right\}}_{=\left(\tilde{u}_{\shortparallel,n+1} - u_{\shortparallel,n}\right)/\Delta z} \right] \end{aligned}\right\}$$

$$\gamma_{n+1} = \sqrt{1 + u^2_{\shortparallel,n+1} + |\vec{p}_{n+1}|^2} \tag{4.18}$$

In the calculations of the acceleration terms $\vec{a}_n$, required for calculation of the electron motion equation in the modified SELFT code, the influence of the space charge effect has also been included and is calculated according to equation (4.19) in which the quantity $\vec{p}_n$ has been calculated according to equation (4.18). This means that in contrast to [Ker96] the calculation of the space charge effect also includes the effect of the non-uniform distribution of the magnetic field. The detailed derivation of this equation can be found in [Ker96, and references therein]. In equation (4.19) the plasma frequency ω_p is calculated using the particle normalized initial velocities $\beta_{\perp,in}$ and $\beta_{\shortparallel,in}$ respectively.

$$\frac{j\omega_p^2}{2\Omega_0}\left(\langle\vec{p}_j\rangle_j - \vec{p}_j\right), \quad \omega_p^2 = \frac{e^2 n_0}{m\varepsilon_0} = \frac{2\pi I_B}{8500\mathrm{A}}\frac{\Omega_0 c}{4\pi R_e \beta_\perp \beta_\shortparallel} \tag{4.19}$$

According to equation (4.18), it has been clear that the non-uniform distribution of the magnetic field has a direct influence on the calculations of the momentum of the particles traveling through the resonator under the influence of the modified RF field profile in the SELFT simulation which is very essential in view of performing investigations of the appearance of dynamic ACI in the output part of the gyrotron cavity. Equation (4.18) has been solved in parallel to equation (4.13) in order to have self-consistent solutions using the modified SELFT code. The inclusion of the effect of the non-uniform distribution of the magnetic field in equation (4.18) is another important modification made in the time-dependent self-consistent differential equations which are responsible for the actual beam-wave interaction calculations considered in the modified SELFT code in comparison to the equations used for the beam-wave interaction reported in [Ker96] where the magnetic field is considered to have a constant value. These modifications are necessary in order to perform investigations towards the occurrence of dynamic ACI phenomena in the output section of the gyrotron resonator where the influence of the change in magnetic field is significant. Using the coupled differential equations (4.18), the movement of each particle has been integrated with the known input parameters for the given geometry. In order to calculate the exchange of power between the electron beam and the RF wave in terms of calculations of the electronic efficiency, i.e. amount of energy transferred from the transverse velocity component of the electron beam to the RF field, equation (4.18) has been solved for each macro-particle (≈ 30). Then the average value of gamma $\langle\gamma_j\rangle_j$ is calculated taken over all the electrons and finally, the averaged value has been used in equation (4.20) in order to get the electronic efficiency, which is given by:

$$\eta_{elec} = \left\{\frac{\gamma_{j,in} - \langle\gamma_j\rangle_j}{\gamma_{j,in} - 1}\right\} \tag{4.20}$$

In principle all the parameters of the electrons, like electron entry location $(R_{e,in,j}, \varphi_{e.in,j})$ (see Fig. 2.1), velocity components $\beta_{\perp,j}, \beta_{\shortparallel,j}$, entry phases relative to the phases of the field

which are described by using a slow phase variable $\lambda_{in,j}$ in the SELFT calculations and the time scale $t_{in,j}$ for each electron which plays a role only for non-stationary simulations, have to be averaged. For the case of stationary as well as non-stationary simulations, the electrons azimuthal angles of incidence have not to be averaged because, in SELFT calculations, the electron beam has been considered as azimuthally symmetric and for the stationary case, in addition to averaging the frequency, one can also consider the averaged electron cyclotron frequency Ω_f to be equal to the mode's oscillating frequency ω_k i.e. $\Omega_f = \omega_k$ so that the time dependencies in the calculations of the acceleration term in the particles motion equation as well as in the excitation term of the RF field profile equation (2.29) can be omitted. This means that all the particles will reach each discrete point of the geometry at the same time step when the RF field calculations are performed at the same discrete points. and thus under this condition the electron beam movement is *quasi-stationary*.
In the case of a non-stationary simulation, the two time scale frames: one for the calculations of the acceleration term required for the electron motion 't''$(\vec{a}_j(z,t'))$ and the other for the calculation of the excitation term required for the RF field profile equation 't''$(\vec{R}_k(z,t'))$, have been considered in the existing SELFT simulations so that it can be easily distinguished between the calculations of the flight time of the electrons and the evolution of the RF field profile as both are calculated in different time frames. In reality, without any assumption, the equation for the motion of each particle is time-dependent and thus in order to have a physically corrected treatment of the non-stationary condition, equation (2.27) can be used without considering the assumption $\partial/\partial t = (u_{\shortparallel}/\gamma)\partial/\partial z$. This is possible in principle but it also increases the computational complexity as compared to the *quasi-stationary* treatment of the electron motion [Ker96]. The typical transit time for electrons through the cavity is around 1 ns i.e. $\tau \approx$1 ns. For several 1000 geometrical discretization steps along the z-coordinate and for the several 10 macro-particles the equations for the electron movement need to be solved, with a time step <1 *ps*, at each simulation time as well as at each discrete point of the geometry which makes a long computational time. If the field profile is assumed to be slowly changing over time, the equations for the electrons need only to be solved once within the time step for the RF field profile. Descriptively, it has been assumed that the excitation term for the RF field profile is nearly constant during the transit time of the electrons. It is possible to identify the greatest possible time step for fast and reliable simulation around 0.01 ns. A second important numerical parameter for reliable simulation is the number of utilized macro-particles. For the simulations in this work, the number of macro-particles considered in the simulation is chosen to be >30. Probably the most crucial numerical parameter is the number of discretization steps for the slow variable phase Λ as described by equation (2.16). A safe choice for the number of discretization steps in this case is the next higher prime number related to the highest azimuthal index m of the modes considered in the multi-mode simulation. This selection reduces the possibility of beating effects and non-physical behaviours.

In SELFT simulations the particle movement is treated as *quasi-stationary* and as a result it is necessary to solve the equation of motion only once after the evolution of the RF field profile. This approach is described as 'fixed field approach'. With this approach the computational complexity is reduced by about $\approx$1000 times depending upon the discretization and number of macro-particle considered in the simulation [Ker96]. Therefore the *quasi-stationary* treatment of the electron motion with equation (2.29), where the condition given by equation (2.28) is already implied, has been used in this work as it allows a much faster as well as sufficiently accurate simulation. It is also necessary that the RF field profile that acts on the electron must be obtained at the same time step at which the electron arrives at the appropriate geometry point. But under the consideration of the 'fixed field approach', the excitation term for the RF field profile must be backdated to a time step at which the particle movement is considered. In the modified SELFT code, this alteration is numerically implemented in the time factor associated with the calculation of the excitation term $\vec{R}_k(z,t)$ required for the evolution of the RF field profile.

It is necessary to mention that the flight time of the electrons through the simulation space τ(several ns) is much higher than the time step $\Delta t(\approx 0.01)$ required for the evolution of the RF field profile. This fact has been considered in the existing SELFT code both in the phases of the RF field at the local positions of the electrons as well as in the acceleration terms for the electron motion. In the existing SELFT code this has been corrected by assuming that the field amplitude remains sufficiently constant over a time but the RF field phase varies harmonically with the difference of the actual oscillation frequency ω_k and the chosen carrier frequency $\omega_{carrier}$ (carrier frequency modulates the physical field down to the slowly varying field $\vec{f}_k(z,t)$). A large time step for the field calculation i.e. $\Delta t > t\_field$ $(= (Q/\omega_k))$ has been taken into consideration if the beating period between this two frequencies $t\_dev = 1/|\omega_k - \omega_{carrier}|$ is sufficiently large. In the case when $\tau > \Delta t$, the phase of $\vec{f}_k(z,t)$ changes considerably during the electron transit. This phase change has been considered in the SELFT code by predicting the field phase during the electron transit and this has been accomplished by assuming harmonic behaviour of the field over short times which has been used to extrapolate the fields through an exponential term and also to backdate the calculated excitation terms from the electron movement. As a consequence with this technique and under the given assumptions, $\tau > \Delta t$ has been handled in the frame of the SELFT code without any error. In the SELFT code for non-stationary multi-mode simulation, the total electric field acting on the electrons is considered to be a vectorial sum of the fields of individual modes and the beating process among these modes. This fact has been maintained in the simulation with many different beating periods which means that in this non-stationary case also the technique of extrapolating field phases over the electron motion along the z-axis, as described above, is mandatory as long as a trajectory code has been used for the simulations. Thus we can say that under the assumption made in the SELFT code (the individual mode fields behave sufficiently harmonic with a sufficient stationary amplitudes-'fixed field approach'), the electrons experience the correct field and the calculated excitation term for the field profiles are correct as well. According to the above discussion it appears that the appearance of the dynamic ACI (see chapter 6) in the uptaper section is neither an artefact due to the violation of the assumptions in the modified SELFT code nor due to numerical reflections but it is due to the cavity oscillations which influence the electron movement. In other words if the ACI phenomenon does not essentially change the relevant excitation term of the RF field profile then it would not be self-sustained but powered by the cavity oscillations. As a result we can assume that the appearance of dynamic ACI phenomena in the simulated results (see chapter 6) are real since the excitation term would still be calculated correctly despite the appearance of the dynamic ACI later in the simulation domain due to electron movement in the uptaper section.

In the SELFT simulations, the time scale required for the calculation of the field profile t has been associated with the calculation of the time scale required for solving the electron motion equation (2.29) i.e. with t' and this has been done by using the following relation [Ker96]:

$$t' = t + \int_{z_{in}}^{z} \frac{1}{v_{\shortparallel,j}} d\zeta = t + \int_{z_{in}}^{z} \frac{\gamma_j}{c u_{\shortparallel,j}} d\zeta \tag{4.21}$$

This means that at each time step for every simulated electron the complex exponential terms having a factor of t present in the calculations of the excitation term as well as acceleration term are required to be calculated, since equation (4.21) depends on the longitudinal velocities of each individual electron which was considered as a constant value for each electron in [Ker96]. Solving the exponential term at each time step is numerically very expensive and thus can be avoided by using the following approximation:

$$t' = t + \left\{ \frac{(z_n - z_{in})}{v_{\shortparallel,j,in}} \right\}_{\text{Time taken to travel from the initial grid point to any particular grid point}} \tag{4.22}$$

This approximation is good enough only with a particular condition i.e. when $v_{\shortparallel,j}$ for each electrons are assumed to be constant which is mentioned in [Ker96] but in this present work $v_{\shortparallel,j}$ for each electron is no longer constant due to the inclusion of the non-uniform distribution of the magnetic field in solving the electron motion equations. The consideration of the longitudinal velocity for each electron in the calculations of t' according to equation (4.22) makes the simulation very slow. Thus in order to have a fast and reliable simulation as close to reality as possible, $v_{\shortparallel,j}$ has been calculated as an average value taken over a number of macro-particles at each discrete point of the geometry i.e.

$$t' = t + \left\{ \frac{(z_n - z_{in})}{\langle u_{\shortparallel}(z) \rangle_{J=\text{No. of macro-particles}}} \right\}_{\text{Time taken to travel from the initial grid point to any particular grid point}} \tag{4.23}$$

This is another modification which is included in the modified SELFT code. The impact of the above approximation (4.23) can easily be checked by varying the average electron cyclotron frequency Ω_f with respect to the average field oscillating frequency ω_k since the term t', present in the exponential term required for the excitation as well as acceleration term calculations in the modified SELFT code, is associated with the difference between these two averaged frequencies. These frequency differences have a maximum frequency separation bandwidth where the simulations have been found to be sufficiently good to match with the correct and reliable results. The results obtained with the above approximation (4.23) are well within the required excitation band and remain acceptable.

In [Ker96] the slow phase variable has been calculated by using the relation shown in equation (2.16) where Ω_f has been considered to be an arbitrary constant averaged reference frequency which has been taken to be very close to the electron cyclotron frequency. This assumption is valid enough when the magnetic field taper is small (≈5-6%) but for the case of large magnetic field taper there is a violation of energy conservation. This means that in the beam-wave interaction calculations there is significant difference between the RF efficiency as well as the electronic efficiency values and also the simulation requires a very small z-step size for getting a good result without any numerical overflows. As a result the computational time is longer. In order to simulate beam-wave interaction phenomena self-consistently in the case of a large tapered magnetic field (≈10-15%) profile it is necessary to modify the average reference frequency so that it should vary slowly with reference to the axial coordinate of the resonator and remains no longer a constant value. As a consequence it avoids the violation of the energy conservation as well as numerical overflows in the simulation results. Simultaneously it can also handle larger z-step size along with the larger taper in the magnetic field profile. In this case the reference frequency has been considered to be varied according to the varying magnetic field profile along the axial coordinate of the resonator and uptaper. Thus the modified version of (2.16) can be represented by:

$$\psi = -\Lambda(t) + \Omega_f(z)t + \zeta \tag{4.24}$$

with z-dependent electron cyclotron frequency.

5 Parametric/validation study of the modified SELFT code

The proper choice of the numerical parameters for non-stationary self-consistent multi-mode calculations for gyrotron cavities, using the modified version of the SELFT code, like time step size (temporal discretization), z-coordinate step size (spatial discretization) and number of electrons distributed at the cavity entrance relative to the field phases will be discussed in this chapter. The figure of merit in our parametric study are the resonance frequencies of the main mode and the parasitic/spurious modes. Recent megawatt class gyrotrons required for fusion plasma heating have reached not only to high power but also require high order operating modes. Thus mode competition calculations for gyrotron interaction get increasingly complex, and as a consequence, the proper choice of numerical parameters is obvious very important for the convergence of the modified SELFT code and simultaneously also required for proper estimation of the physical behaviour of the gyrotron interaction. Speaking for the bandwidth of included modes, parameter changes during the calculation must be considered. Typically, in gyrotron startup simulations the cyclotron frequency decreases with increasing voltage. Therefore the frequency band must be expanded to ± 10 GHz around the working mode. Having determined the list of modes, it is a save choice to set the time step size to less than half of the inverse of bandwidth. It is also necessary to have proper azimuthal discretisation that helps to avoid artificial coupling between some modes. In this chapter a parametric study of the modified SELFT code has been taken into account in the frame of slow variable codes for time-dependent, self-consistent multi-mode calculations.

The electron beam has to be modelled through a number of discrete macro-particles with the desired distribution functions over radius, velocities, entry phase and azimuthal position. In the KIT SELFT code, for the case of a single-mode simulation, entry phase and azimuthal position are equivalent and thus there is no need for azimuthal discretization. In the multi-mode simulation case, the beating phase between modes destroys this equivalence. But in the modified SELFT code it is still not necessary to do a full azimuthal discretization, since the beating phase between each pair of modes repeats itself over the azimuth. The azimuthal discretization can be sparse, but must not be synchronized with any mode beating phases.

In non-stationary single-mode simulations primarily three parameters are discretized namely the z-coordinate, simulation time and the distribution of all electrons at the entry plane of the resonator relative to the input field phase Λ_{in} (for details see [Ker96]) but for the case of a multi-mode simulation using the SELFT code the azimuthal angle of incidence of the electrons φ_e (see Fig. 2.1) is also discretized in consideration with the velocity dispersion of the electron beam. In the existing SELFT code the velocity distribution function has been modelled by using standard deviation with three sample points but for the energy distribution of the electrons after the beam-wave interaction in the depressed collector, it is required to consider more sample points (e.g. seven) [Ker96].

For the discretization in the z-coordinate, in order to be sure that the oscillatory behaviour of the RF field in the region of the output is reproduced accurately, a discretization step width $\Delta z<\lambda_0/10$ is sufficient. Whatever numerical method is used to integrate the differential equations, accuracy will be poor if $\Delta z<\lambda_0/10$ [Bor91]. For a further discussion of the effect of step width in integrating the differential equations see [PTV+92] whereas the averaging over φ_e has already been discussed in [Ker96]. In the SELFT code the averaging over φ_e has been done with 7-10 averaging steps and this is sufficient for the proper estimation of the beam-wave interaction as according to the experience gained with the existing SELFT code; there is no substantial change in the simulation results with the increase in the number of averaging steps.

Due to the present modifications made in the differential equations responsible for the beam-wave interaction calculations (see chapter 4) in the modified SELFT code, the convergence study due to the discretizations in the simulation time as well as the number of electrons required for the discretization of the entry phase of the field Λ_{in} is necessary. This is important due to the presence of the synchronization between the flight time of the electrons and the phase change of the field profile. In the SELFT code the common choice on the number of electrons required for the discretization of the Λ_{in} is around 30 (i.e. $n_e \approx 30$). For stationary multi-mode simulations, this number is not sufficient, because when the ensemble of electrons (macro-particles) move towards increasing z values, the distribution of the electrons over Λ_{in} cannot be well represented since most of the electrons are bunched in the period of maximum energy radiation. Fig. 5.1 represents the phase bunching of the macro-particles as they propagate towards increasing z-values.

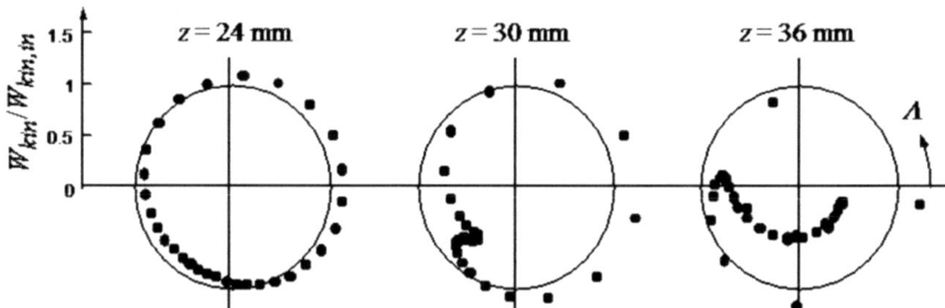

Figure 5.1: Phase bunching of the particles in the plane along z-coordinate (adopted from [Ker96] for illustration purpose).

In order to solve the problem the SELFT code adopts a possible solution to distribute the macro-particles at regular intervals of λ_{in} so that the distribution curves over the $W_{in} - \Lambda$ plane are again sampled uniformly [Ker96].

Whereas on the other hand in non-stationary multi-mode simulations the above problem is mitigated because in this case the phase of the field profile at the resonator input changes over time, and thus the relative entry phase. A moderately poor distribution of the electrons in a particular time step that produces primarily fluctuations in the simulated results gets averaged out overall several time steps and as a result the simulated results are sufficiently good having few numerical noises.

It is also required to mention that a too coarse time discretization step leads to deviations in the field profile calculations, particularly when the excitation term ($\vec{R}_k$) is found to be large enough. While on the other hand, due to the assumption of the quasi-stationary movement of the macro-particles in the SELFT code, the particle motion is not affected by the time discretization step and as a result the energy delivered by the bunched electrons to the RF field profile is then calculated correctly even though there is an improper discretization in the time. For more details see [Ker96]. The ambiguity due to the time discretization step size may be expressed as a violation of energy conservation. which results in a significant difference between the two calculated efficiencies (RF as well as electronic efficiencies). This difference in efficiencies reflects that the bunched electrons are not able to deliver their energy to the RF field properly. Thus in the KIT SELFT code the measure (or amount) of the violation of the energy conservation is determined by how large the time step size has been chosen. Therefore it is necessary to use an optimized value of the time discretization step size in order to estimate the proper gyrotron beam-wave interactions. For the case of multi-mode simulations due to the presence of beating frequencies, in the KIT SELFT code the time discretization step size is generally not allowed to be greater than half of the beat period. The time step may be further increased, but it must at least be small enough to describe beating frequencies between any active modes [KAB+08]. It is required to mention that the onset of mode competition occurs correctly when the discretization time step size is about 1/4 of the beat period between the competing modes so that the mode eigen frequencies can be of the order of the excitation band [Ker96, Nus04].

In the existing time-dependent self-consistent SELFT code, capable for multi-mode simulations, the proposed modifications (see chapter 4) have been implemented in order to investigate the influence of a non-uniform distribution of the magnetic field on the occurrence of dynamic ACI in the uptaper section of the cavity geometry. In these modifications mainly the terms containing $dB_{R,z}/dz$ have been included in the differential equations responsible for beam-wave interaction. Due to the proposed modifications (see chapter 4), it is necessary to perform parametric studies to investigate the effect of the numerical parameters on the gyrotron beam-wave interaction calculations. In the present work the study is based on numerical experiments performed using the modified version of the SELFT code in order to confirm the convergence of the simulated results due to the modified differential equations. In this study convergence of the parasitic frequency has been observed while changing the various above mentioned numerical parameters. In this parametric study, the $TE_{-28,8}$ mode 140 GHz W7-X gyrotron has been chosen as an example in order to perform investigations on the effect of numerical parameters on the parasitic/spurious oscillations which can be suspected to be due to the occurrence of dynamic ACI. In this calculation, the beam parameters are: $V_{beam} = 81.8$ kV, $I_{beam} = 43.2$ A, $\alpha = 1.18$, $B_0 = 5.615$ T and $R_{beam} = 10.17$ mm which are typical experimental beam parameters where a parasitic oscillation at 130.30 GHz has been measured (see chapter 6). This chapter has been dedicated to parametric single-mode studies with respect to the following parameters:

- Parametric study with respect to temporal discritization.
- Parametric study with respect to spatial discritization.
- Parametric study with respect to number of electrons for azimuthal phase discritization.

5.1 Effect of temporal discritization step size on the parasitic frequency oscillations

In this section the effect of the variations of the temporal discretization step size on the simulated frequency of the parasitic oscillations are presented. The aim of this work is to determine the range of Δt in which the simulated results are found to be reasonable. In this case time-dependent self-consistent non-stationary single-mode simulations have been performed using the above mentioned beam parameters in the modified version of the SELFT code. In this calculations the step size for the spatial discretization as well as the number of macro-particles required for the initial azimuthal phase discretization are considered to be 0.1 mm and 30 respectively. The RF spectrum of the simulation results, considering different time discrietization step sizes is plotted in Fig. 5.2. One can observe that the solutions, obtained with the modified version of the SELFT code, converge to the experimentally obtained frequency of 130.30 GHz as the value of Δt reduces. The zoomed picture of the encircled region of Fig. 5.2a is plotted for clear visualization purpose in Fig. 5.2b.
Also Fig. 5.3 shows that by decreasing the time discretization step Δt the solution converges to a parasitic frequency (130.4 GHz) which is found to be very close to the experimentally obtained value (i.e. $\approx$130.30 GHz). This convergence shows that the implementation of the proposed modifications in the actual time dependent differential equations (see details in chapter 4) for the estimation of dynamic ACI phenomena is found to be correct with respect to the time discretization step size . Thus for fast and reliable simulations, the time discretization step size Δt has been taken to be 0.01ns in all the further simulations reported in this thesis performed by the modified version of the SELFT code (for simulation results see chapter 6). In this case the theoretical calculated parasitic frequency is 130.8 GHz.

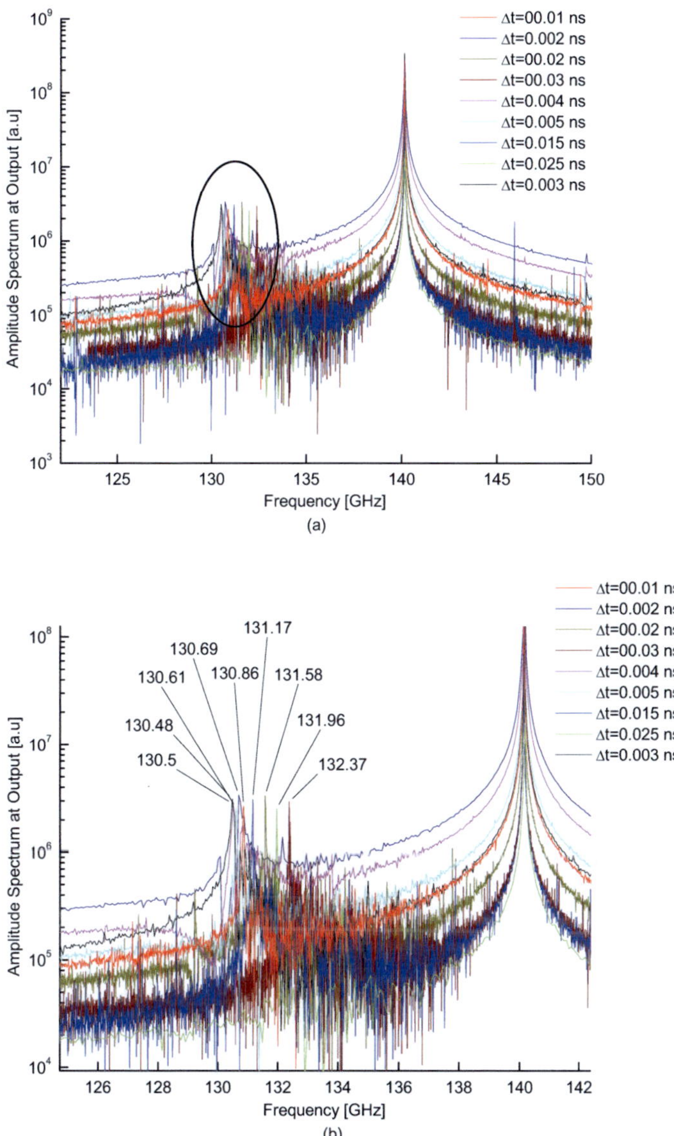

Figure 5.2: (a) Study of convergence with respect to the time discretization step size (Δt), (b) zoomed picture of the encircled portion of (a) where the variation of the parasitic frequencies according to the change in is shown explicitly for clear visualization.

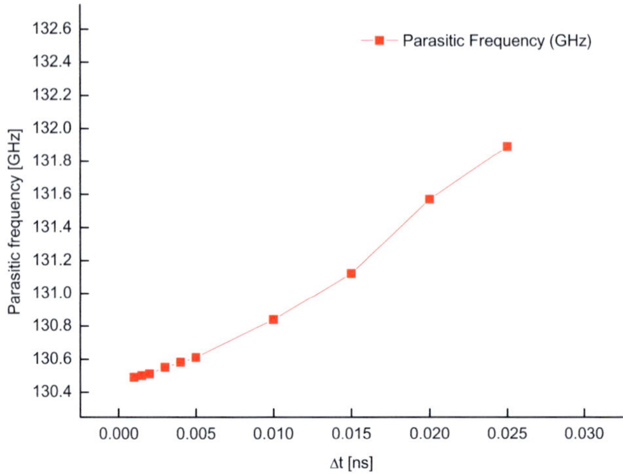

Figure 5.3: Graphical representation of the convergence of the solution with the decreasing time discretization step size (Δt).

5.2 Effect of spatial discretization step size on the parasitic frequency oscillation

In this section investigations on the effect of the variations of the spatial discretization step size on the modified differential equations implemented in the modified version of the SELFT code required for time-dependent self-consistent multi-mode simulations are presented. In this study again the figure of merit is the simulated frequency of the parasitic oscillation.The temporal discretization step size Δt and the number of macro-particles required for the initial phase discretization are considered to be 0.01 ns and 30 respectively. Non-stationary single-mode simulations are performed using the above mentioned beam parameters in the modified version of the SELFT code with different spatial discretization sizes.

The simulated RF spectrum plot obtained with different space discritization step sizes is presented in Fig. 5.4. The zoomed picture of the encircled portion of Fig. 5.4a is also presented for the clear visualization in Fig. 5.4b in which one can observe that there are two sets of parasitic oscillations having different frequency range. Each set of parasitic frequency belongs to different values of Δz. This sudden jump in the parasitic frequencies to a lower frequency value (see Fig. 5.4) is attributed to the fact that it is due to the long-line effect [BJ90]. This phenomenon can be addressed by using Rieke diagrams - a diagram that gives real and imaginary part of the reflection coefficient along the contour of the constant output

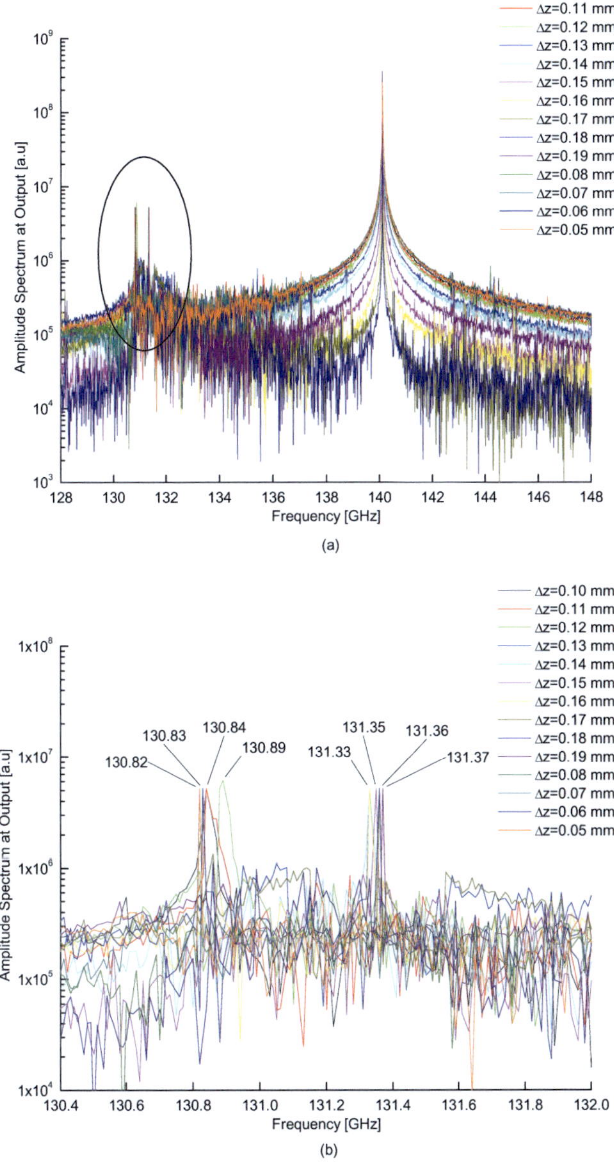

Figure 5.4: (a) Study of convergence with respect to the space discretization step size (Δz), (b) zoomed picture of the encircled portion of (a) where the variation of the parasitic frequencies according to the change in are shown explicitly for clear visualization.

power and/or frequency [Bor01]. The Rieke diagram provides a quick method for determining whether or not a solution exists for the parameters and reflection coefficient of interest, and to determine approximately the value of the frequency to which the specific solution corresponds. The phase of the electric field in the cavity is generally affected by the long-line effect and is poorly known since it is difficult to measure [BJ90]. In this condition the phase of the reflection coefficient at the tapered output of the cavity plays an important role. The effect of reflections on gyrotron operation depends not only on their magnitude but also their phases. The change in phase is determined by the changing distance of the resonator output section over a certain distance corresponding to a phase shift of 0 to 2π. The influence of the phase change of the reflection coefficient on the calculated parasitic frequency has been depicted in the graphical representation of the parasitic frequencies along with the Δz variations where the frequency jumps suddenly after a certain value of Δz (see Fig. 5.5). In the KIT SELFT code, the reflection coefficient is calculated according to equation (4.8) where Δz has a direct influence on the phase of the reflection coefficient which in turn has an influence on the long-line effect. From Fig. 5.5 it is also clear that the solutions, obtained with the modified version of the SELFT code, converge close to 130.8 GHz (see section 5.2) and tend to stabilize as long as the value of lies within the range 0.05-0.12 mm. On the other hand, the solutions jump to higher frequency values as soon as lies between 0.13-0.19 mm range due to the long-line effect.

From equation (4.8) it can be seen that the change in Δz has a direct influence on the numerical calculations of the output reflection coefficient (both in magnitude and phase). Due to the change in Δz values, the numerical value of the reflection coefficient shifts to another contour of constant frequency, which corresponds to a higher frequency, in the Rieke diagram plot and this reflected in Fig. 5.5. However from Fig. 5.5, it is clear that due to the presence of stable solutions, having less divergences, in terms of the parasitic oscillation frequencies in a particular range of Δz shows that the proposed modifications in the time-dependent differential equations (see details in chapter 4) for the estimation of dynamic ACI are validated against the space discretization step Δz. For fast and reliable simulations, the space discretization step size Δz has been considered to be 0.10 mm in all the further simulations performed by using the modified version of the SELFT code which are reported in this thesis (for simulation results see chapter 6).

5.3 Effect of number of electrons required for azimuthal phase discretization on the parasitic frequency oscillations

In this section the effect of the number of electrons required for the azimuthal discretization of the electron beam which, in the existing SELFT code, appears as an azimuthal phase averaging among modes, has been validated against the modifications made in the required differential equations. In this validation work the convergence of the parasitic oscillating frequency has been taken into consideration. Single-mode non-stationary self-consistent simulations, using

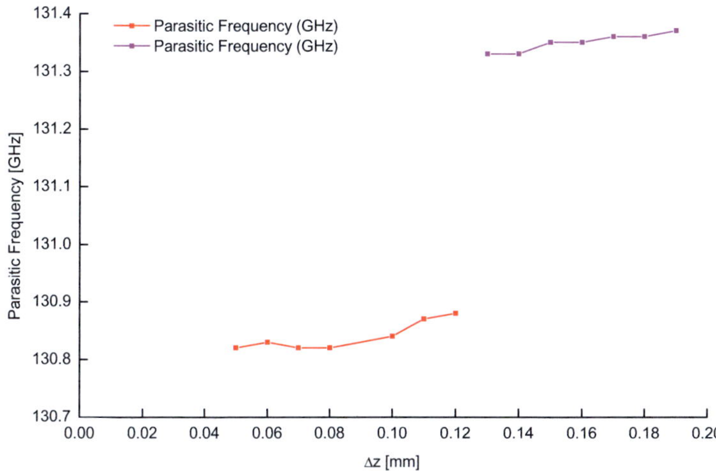

Figure 5.5: Graphical representation of the convergence of the solution with decreasing space discretization step size (Δz). The plot represents two different set of parasitic frequencies (shown by different colours) at the two different set of values.

the above mentioned beam parameters, are performed in order to investigate the convergence of the solution due to the modified differential equations in terms of parasitic oscillating frequency. RF spectra of the simulated results for different values of number of electrons are plotted in Fig. 5.6. In this study the temporal step Δt and the spatial discretization step size Δz are considered to be 0.01 ns and 0.10 mm respectively in the modified version of the SELFT code. An expanded view of the encircled portion of Fig. 5.6a is presented in Fig. 5.6b in order to have a clear visualization of the variation of the parasitic frequency over the number of electron required for the initial azimuthal phase discretization.

From the above zoomed picture (see Fig. 5.6b) one can observe that the parasitic oscillation frequency converges close to 130.8 GHz (see section 5.2). This is the case when the number of considered macro-particles in the modified SELFT code is greater than 30. The parasitic oscillating frequencies are also plotted graphically against the different number of macro-particles required for the initial azimuthal phase discretization and are shown in Fig. 5.7. From Fig. 5.7 it also can be seen that the solutions, in terms of the parasitic frequencies, converge to 130.8 GHz (see section 5.2) and remain stable. It also shows that there is no effect on the simulated results as long as the number of macro-particles is greater than or equal to 30. Thus the above numerical experiment confirms that the proposed modifications in the time dependent differential equations (see details in chapter 4) of the modified version of the

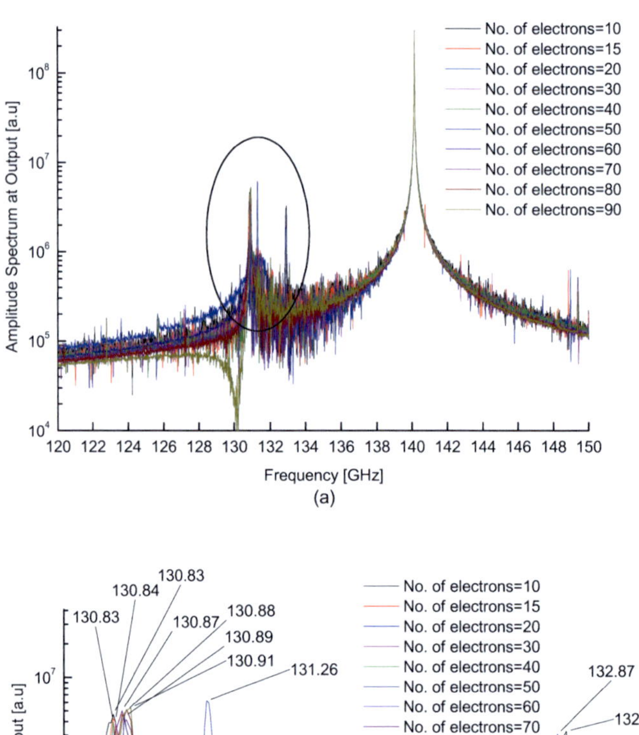

Figure 5.6: (a) Study of convergence with respect to the number of electrons required for azimuthal phase discretization.(b) zoomed picture of the encircled portion of (a) where variation in the parasitic frequencies according to the change in the number of electrons are shown explicitly for clear visualization.

SELFT code in order to estimate dynamic ACI phenomena are validated against the number of macro-particles required for the initial azimuthal phase discretization. For fast and reliable simulations, the number of macro-particles have been considered to be 30 in all the further simulations performed using the modified version of the SELFT code in order to investigate dynamic ACI in the presence of a non-uniform distribution of the magnetic field (for simulation results see chapter 6). While correct temporal discretization as well as spatial discretization

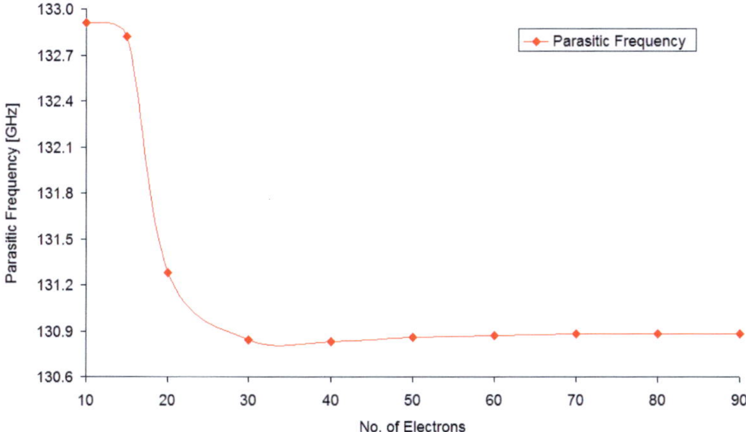

Figure 5.7: Graphical representation of the convergence of the modified version of the SELFT simulated results with the increasing number of macro-particles required for the azimuthal phase discretization.

step sizes are crucial for obtaining reliable simulation results, as discussed above, the number of macro-particles is critical when stationary self-consistent simulations are required [DN92] but turns out to be less sensitive in time dependent non-stationary simulations. On the other hand it depends on the number of modes considered in the multi-mode simulations. The bunching process, which collects macro-particles in a small range of the slow variables phase, is highly desired but leaves only a few macro-particles in a broad region of phase. Macro-particles remaining in such rarified ranges are not representative any more but can have a strong influence depending on their starting positions. In the case of non-stationary simulations the above effects are balanced by the fact that, over time, the influence of such particles is averaged out while for the case of stationary simulations even 300 macro-particles may be not enough for reliable simulation results [KAB+08].

It is required to mention that in the modified SELFT code, the simulation model accounts for the azimuthal structure of each mode $TE_{-m,p}$ analytically but for the multi-mode calculations the phase difference between two modes of different m changes over the azimuthal angle φ (see Fig. 2.1). Since these beating phases repeat over φ, the KIT SELFT code uses only a few

discretization steps n_φ to cover a representative sample of phase differences but not to fully describe the mode structure over φ. In the KIT SELFT code when two modes with azimuthal indices $m1 \neq m2$ fulfil the condition $(m1-m2)/n_\varphi =$ integer, an equidistant azimuthal steps (equal to n_φ) will calculate identical azimuthal phase differences [Ker96]. Generally for reliable simulations using the KIT SELFT code, n_φ can be set to the next higher prime number above $2m_{max}$ but less than the $10m_{max}$ [KAB+08] (m_{max} stands for the maximum azimuthal mode index present in the multi-mode simulations).
A perfect reproduction of the experimental result may not be expected, taking into account realistic tolerances on the one hand, and remaining numerical uncertainties on the other hand. So we got ≈130.8 GHz instead of the experimentally observed 130.3 GHz (see section 5.2).

5.4 Validation of equation of motion solved by PC method against the piecewise linerization method

In order to validate the influence of the inclusion of the term containing the non-uniform distribution of the magnetic field in the electron momentum and axial velocity differential equations (see detail in chapter 4), a second order predictor and corrector method (PC method) has been used to solve the above mentioned differential equations. In this section the results obtained by using the PC method have been compared with the results obtained by an analytical calculation approach i.e. by a piecewise linearized approximated solution.
For the sake of simplicity, the acceleration term $\vec{a}$ (see equation 2.29a) in the electron momentum equation has been set to zero. This is because at the starting point in the simulation domain (resonator cavity) the initial RF field has no influence on the electron motion. If the vector $\vec{p}$ is represented in the complex plane, equation (2.29a) without acceleration term $\vec{a}$ can be separated into real (P_R) as well as imaginary part (P_I) and recasted in the form which is described below.
Let us define

$$A = \frac{cu_{\shortparallel}}{\gamma}; \quad B = \left\{ \frac{cu_{\shortparallel}}{\gamma} \cdot \frac{1}{2B_{R,z}} \frac{dB_{R,z}}{dz} \right\}; \quad \Omega^* = \left(\Omega_f - \Omega_0/\gamma\right)$$

Equation (2.29a) without acceleration term $\vec{a}$ can be written as:

$$P_R' + jP_I' = \left\{ \frac{B}{A} - \frac{\Omega^*}{A} \right\} (P_R + jP_I)$$

Separating real and imaginary part we get:

$$\begin{aligned} P_R' &= \frac{B}{A} P_R + \frac{\Omega^*}{A} P_I \\ P_I' &= \frac{B}{A} P_I - \frac{\Omega^*}{A} P_R \end{aligned} \tag{5.1}$$

where P'_R, P'_I are the spatial derivatives of the real and imaginary part of the vector $\vec{p}$ respectively. In solving equation (5.1), the $u_{\shortparallel}$ values have been obtained by solving equation (2.29b). Each separate part of equation (5.1) has been solved separately by using the PC method and the results ($|p| = \sqrt{p_R^2 + p_I^2}$) are plotted with respect to the axial coordinate in Fig. 5.8. Similarly equation (2.29b) has been solved using a PC numerical method in which the values of $u_\perp$ have been obtained from equation (2.29a) (without accelerating term) and the results are plotted in Fig. 5.8 along with the axial coordinate.

Another way to see the equation (2.29a) (without accelerating term) in terms of the first order homogeneous differential equation which involves only the first derivative of a function and the function itself, with constant only as a multiplier and is given by:

$$\vec{p}' + \vec{A}\vec{P} = 0 \tag{5.2}$$

where $\vec{A}$ is a complex quantity, function of $u_{\shortparallel}, \gamma, \Omega_f, \Omega_0$ and $B_{R,z}$, is defined by

$$\vec{A} = \frac{\left[j\left(\Omega_f - \Omega_0/\gamma\right) - \left(\frac{cu_{\shortparallel}}{\gamma} \cdot \frac{1}{2B_{R,z}} \frac{dB_{R,z}}{dz} \right) \right]}{cu_{\shortparallel}/\gamma}$$

A formal solution of equation (5.2) can be written as:

$$\vec{p}_{n+1} = \vec{p}_n e^{-\vec{A}\Delta z} \tag{5.3}$$

where $\vec{A}$ is any suitable approximation of $A(z)$. Similarly a piecewise linearized solution for equation (2.29b) can be obtained simply intregating equation (2.29b) from z_n to z_{n+1} and can be written as: :

$$u_{\shortparallel}^{n+1} = \sqrt{(u_{\shortparallel}^{n+1})^2 - \vec{u}_{\shortmid}^2 B^* \Delta z} \tag{5.4}$$

where $\vec{u}_\perp$ and B^* are the suitable approximation of $u_\perp(z)$ and $\frac{1}{B_{R,z}} \frac{dB_{R,z}}{dz}$ respectively.

Fig. 5.8 shows the solutions for the homogeneous electron momentum equation as well as electron axial velocity equation by using the two methods mentioned above. One can see that there is very good agreement between numerically obtained results and the results calculated using the piecewise linerized approach. This confirms that the PC method can be used to solve the above differential equations in the modified SELFT code.
The above numerical experiment confirms that the inclusion of the term containing the non-uniform distribution of the magnetic field in the electron motion equations, both in axial as well as transverse direction, can be implemented in the actual time-dependent differential equations responsible for gyrotron beam-wave interaction. It also confirms that the second

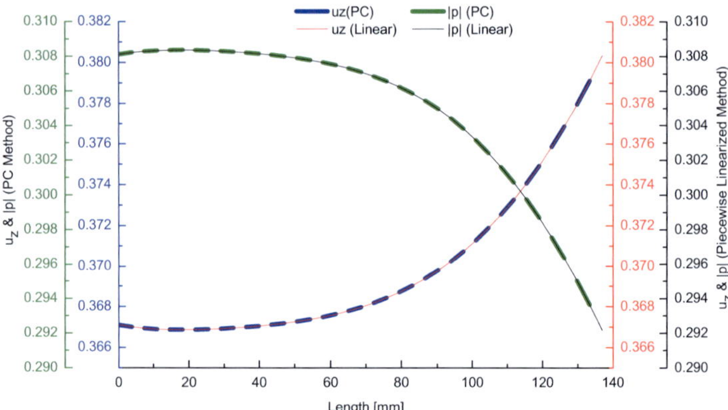

Figure 5.8: Solution of the electron momentum and axial velocity equations using a second order PC method and a piecewise linearized method (analytical approach).

order numerical scheme i.e. Predictor-Corrector method is one of the most suitable numerical methods to solve numerically the actual differential equations of the electron motion. In the modified version of the SELFT code, the Predictor-Corrector method has been implemented (see detail in chapter 4).

6 Dynamic ACI study with the modified SELFT code

In this chapter a dynamic ACI study with the modified version of the SELFT code is presented using four different gyrotron configurations (two conventional cavity gyrotrons and two coaxial gyrotrons). Furthermore in order to verify the implementation of the term containing $dB_{R,z}/dz$ in the differential equations, which are responsible for the beam-wave interaction calculations, the simulated results for the case of the 140 GHz, $TE_{-28,8}$ mode gyrotron developed for the W7-X Stellarator have been compared with the available experimental results with respect to the parasitic/spurious oscillations present in the uptaper section of the resonator. In this study the non-uniform distribution of the magnetic field has been taken into account in contrast to the results described in chapter 3 where constant magnetic field has been used in the differential equations for beam-wave interaction calculations [Ker96]. This chapter has been dedicated to numerical simulations for the following four example gyrotron configurations with the intension to investigate the presence of dynamic ACI in the output section of the gyrotron resonator:

- 140 GHz $TE_{-28,8}$ mode 1 MW conventional cavity gyrotron.
- 170 GHz $TE_{-32,9}$ mode 1 MW conventional cavity gyrotron.
- 170 GHz $TE_{-34,19}$ mode 2 MW coaxial cavity gyrotron.
- 240 GHz $TE_{-55,29}$ mode 2 MW coaxial cavity gyrotron.

In these numerical simulations, with high electron beam coupling and within a bandwidth of ± 10 GHz around the oscillating frequency of the main operating mode in the cavity have been taken into account. The coupling limit was chosen to be 70% of the operating mode, since modes operating in gyrotron cavities tend to suppress other modes with even only slightly less coupling. The mode selection has been determined using the program called "SCNCHI" [Ker96].

The simulated results presented in this chapter may be viewed as a validation of all the proposed modifications implemented in the SELFT code for the investigation of dynamic ACI phenomena in the uptaper section of gyrotron resonators.

6.1 140 GHz 1 MW $TE_{-28,8}$ mode conventional cavity gyrotron

In this section the investigation of dynamic ACI in the uptaper section of the resonator of the 140 GHz W7-X gyrotron developed for the W7-X Stellarator is described. In these calculations, the beam parameters are: $V_{beam} = 81.0$ kV, $I_{beam} = 39.0$ A, $\alpha = 1.13$, $B_0 = 5.59$ T and $R_{beam} =$ 10.10 mm. These are parameters for which calculations with a constant magnetic field show no relevant ACI. Multi-mode calculations have been performed, up to 2400 ns simulation time, under the consideration of all the proposed modifications, described in chapter 4, implemented in the modified version of the KIT SELFT code which includes the influence of the non-uniform distribution of the magnetic field profile.

In this particular case, initially with the magnetic field profile obtained from the external code ESRAY existing at KIT [Ill97], it has been found that the $dB_{R,z}/dz$ profile, required for the beam-wave interaction calculations, calculated with a second order central difference scheme, as described in equation (4.16), showed numerical oscillations which may affect the SELFT calculations (see Fig. 6.1). In order to tackle this problem, $B_{R,z}$ was first best fitted with a fifth order polynomial curve and then the first derivative was analytically calculated. Equations for the normalized (with respect to the maximum value) $B_{R,z}$ and $dB_{R,z}/dz$ are shown below:

$$B_{R,z} = 4.851e^{-17}z^5 - 1.883e^{-13}z^4 + 1.904e^{-10}z^3 - 1.92e^{-07}z^2 + 2.493e^{-05}z + 9.981e^{-01} \tag{6.1}$$

$$dB_{R,z}/dz = 24.255z^4 - 7.532z^3 + 5.712z^2 - 2.184z + 2.493$$

where z is the axial coordinate. Equation (6.1) is obtained by using polynomial least-squares fitting technique [PTV+92]. This corresponds to the magnetic field profile labelled as $B_{R,1}(z)$. A steeper magnetic field profile namely $B_{R,1}(z)$ has been proposed for comparison reasons. Both profiles are shown in Fig. 6.2. Some technical references [SN09, SNA10] indicate that the modification in the axial distribution of the magnetic field profile can result in a reduction of dynamic ACI effects and this should be verified by performing advanced simulations using the modified version of the SELFT code with multi-mode simulations applying the two different axial magnetic filed profiles $B_{R,1}(z)$ and $B_{R,2}(z)$.

In order to estimate the impact of the term $dB_{R,z}/dz$ in the discretized differential equations described in chapter 4 in terms of dynamic ACI phenomena, numerical experiments have been performed using the two different versions of the SELFT code i.e. SELFT with uniform magnetic field and the modified SELFT with non-uniform magnetic field. Comparison of the electric field profiles are shown in the following figures Fig. 6.3 and Fig. 6.4 respectively.

The electric field profiles obtained in the case when the magnetic field has been considered as constant with a value which corresponds to the maximum value of the $B_{R,1}(z)$ magnetic field ($B_{R,1}(z) = 5.59$T) profile (shown by solid line) and also when considered as a non-uniform distribution of the $B_{R,1}(z)$ magnetic field profile (shown by line + symbol) are presented in Fig. 6.3. Here only those modes are shown which are relevant for our comparisons. This means that only those modes are shown in the respective plots which are well above the noise level. These electric field profiles which are recorded at simulation time 2400 ns show that considering a non-uniform magnetic field profile, in addition to the operating mode $TE_{-28,8}$, an undesired

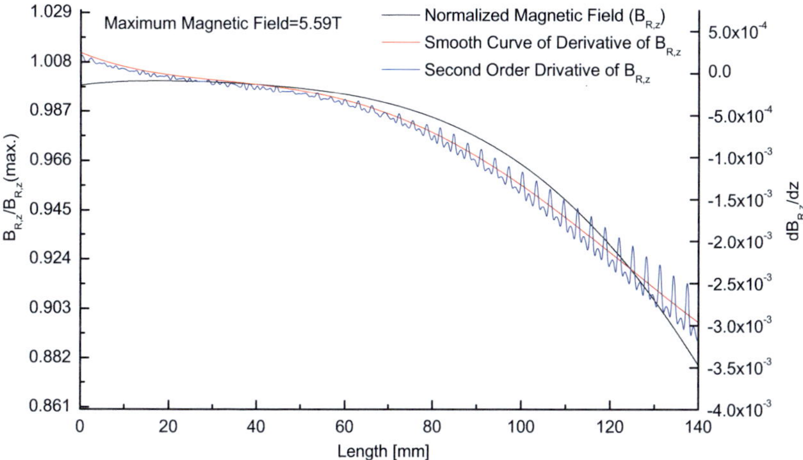

Figure 6.1: Magnetic field $B_{R,1}(z)$ and its derivative along the z axis.

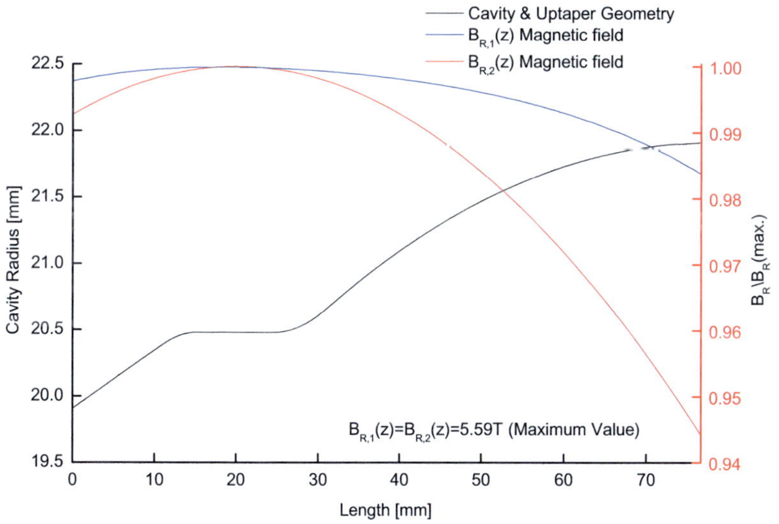

Figure 6.2: Two magnetic field profiles along the z axis.

parasitic mode $TE_{-26,8}$ has also been oscillating in the cavity uptaper region, with a significant power level of ≈7.8 kW while, on the contrary, this does not happen if $B_{R,1}(z)$ is constant. This undesired oscillation can be addressed to the fact that the magnetic field towards the geometry output is significantly different from the magnetic field value at the mid of the cavity and can be considered as the dynamic ACI effect because interaction after the cavity leads to a dynamic amplitude modulation with $TE_{-26,8}$ at lower frequency. One can also observe from Fig. 6.3 that, in both the cases, the field profiles for the desired operating mode are found to be almost identical. Similarly self-consistent multi-mode simulations are also performed using a uniform value of the $B_{R,2}(z)$ magnetic field profile which corresponds to it's maximum value as well as using a non-uniform $B_{R,2}(z)$ magnetic field profile. The results are shown in Fig. 6.4 and they reveal that due to the steeper magnetic field profile $B_{R,2}(z)$, the impact of the undesired parasitic mode $TE_{-26,8}$ as an dynamic ACI effect is significantly reduced.

Fig. 6.5 summarizes the results shown in Fig. 6.3 and Fig. 6.4: a steeper magnetic field profile can suppress the parasitic mode $TE_{-26,8}$. However, apparently from Fig. 6.5 one can also observe that the $TE_{-28,9}$ mode appears as a parasitic mode, contributing as a dynamic ACI phenomenon due to amplitude modulation with a lower frequency at the cavity uptaper section. This can be more evident in the RF spectrum plot (see Fig. 6.9). The reason behind the appearance of this unwanted parasitic mode $TE_{-28,9}$ has been predicted due to the significant change in the beam-wave coupling factor [Ker96, KBT04] due to the large variation in the $B_{R,2}(z)$ magnetic field profile.

Gyrotron interaction efficiency and RF efficiency have also been calculated for the cases when $B_{R,z}1(z)$ is constant as well as non-uniform magnetic field distribution in order to determine the dynamical behaviour of the gyrotron under the considerations of all the above proposed modifications made in the SELFT code. Dynamics of the gyrotron i.e. change in energy stored in the cavity with respect to time, can be determined with the energy balance equation [KBT04]:

$$\frac{dW}{dt} = \eta(t)V(t)I(t) - \frac{\omega W}{Q} \tag{6.2}$$

where W is the energy stored in the cavity, Q is the quality factor, $\eta(t)$ is the total efficiency as a function of time, $V(t)$ is the beam voltage as a function of time and $I(t)$ is the beam current as a function of time. From Fig. 6.6 it can be seen that due to this varying magnetic field profile there are some differences in the efficiencies during the transient start-up phase but both cases reach the same efficiency value (≈19%) in the steady state phase. This low value of efficiency is due to the specific operation parameters used for the present study. The efficiency plot Fig. 6.6 also describes the presence of energy conservation in both cases. This is because due to having a small difference (≈1%) between RF efficiency as well as interaction efficiency for each case in the steady state phase i.e. after ≈1200 ns.

The RF spectrum plot, corresponding to the simulated electric field profiles (see Fig. 6.3), calculated using $B_{R,1}(z)$ equal to the constant value as well as using $B_{R,1}(z)$ equal to the

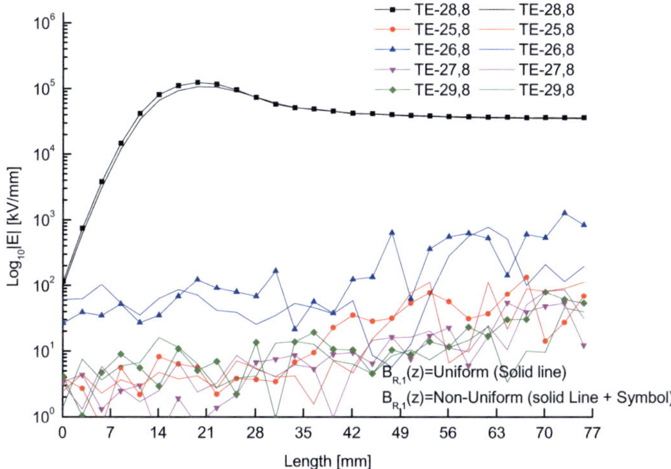

Figure 6.3: Comparison between electric field profiles obtained in multi-mode calculation with $B_{R,1}(z)$ = uniform as well as with $B_{R,1}(z)$ = non-uniform.

non-uniform distribution of the magnetic field profile in the modified version of the SELFT code is presented in Fig. 6.7. The RF spectrum has been recorded at the axial coordinate $z =$ 100.0 mm and at the simulation time 2274 ns for W7-X 140 GHz gyrotron where only those modes are shown which are relevant for our comparisons.

In Fig. 6.7 the dashed line indicates the RF spectrum of the electric field profile calculated with the non-uniform magnetic field profile $B_{R,1}(z)$ while the solid line indicates the RF spectrum of the electric field profile calculated with the uniform magnetic field profile $B_{R,1}(z)$. By taking the zoom picture of the encircled region, shown in Fig. 6.7a, one can see that in addition to the desired cavity operating mode at $\approx$140 GHz, another parasitic modes is also oscillating at $\approx$130.21 GHz ($TE_{-26,8}$ mode) which is due to the dynamic ACI phenomenon. This has been due to the consideration of the non-uniform distribution of the magnetic field profile $B_{R,1}(z)$ in the differential equations (2.8 and 2.29). On the other hand, SELFT calculations with uniform magnetic field $B_{R,1}(z)$ throughout the simulation domain show that beside the desired operating mode other parasitic oscillating modes do not have any significant effect (no dynamic ACI). These calculations clearly confirm that the presence of dynamic ACI phenomena in the uptaper region of the gyrotron cavity can be depicted if the non-uniform distribution of the magnetic field profile is explicitly considered in the differential equations used for the gyrotron beam-wave interaction calculations. Similarly the RF spectrum of the electric field profiles (see Fig. 6.5), recorded at $z = 99.90$ mm and at 2274 ns simulation time have been plotted in Fig. 6.9

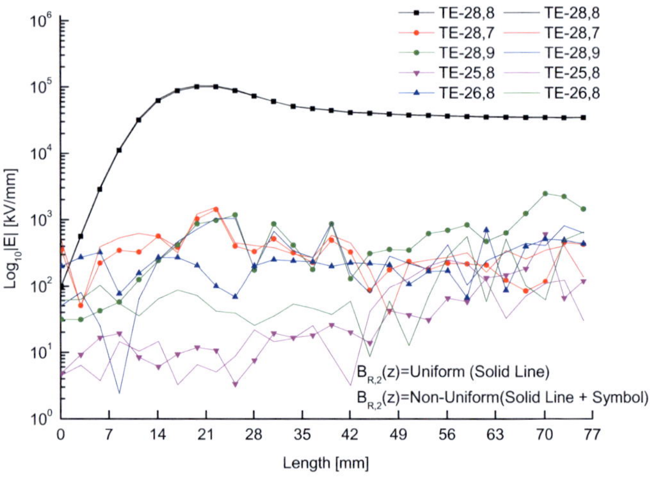

Figure 6.4: Comparison between electric field profiles obtained in multi-mode calculation with $B_{R,2}(z)$ = uniform as well as with $B_{R,2}(z)$ = non-uniform.

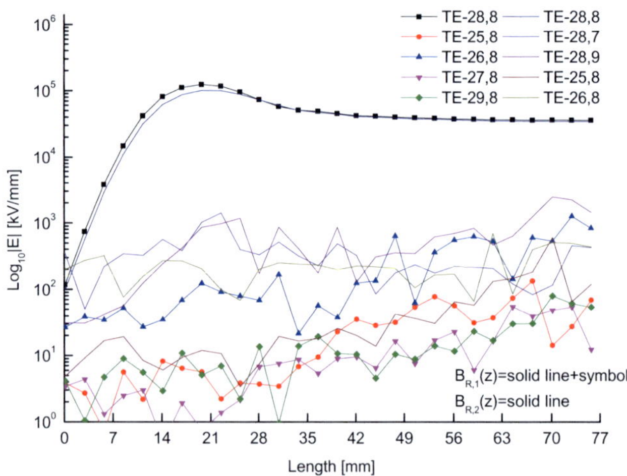

Figure 6.5: Comparison between electric field profiles obtained in multi-mode calculation using the modified SELFT code with non-uniform $B_{R,1}(z)$ and $B_{R,2}(z)$ magnetic fields profiles.

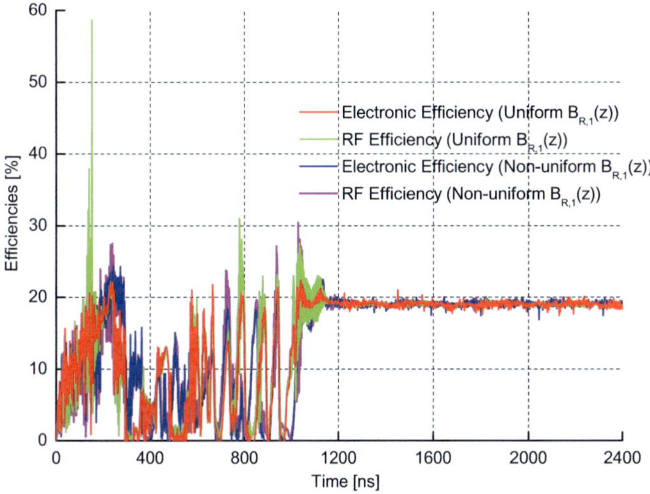

Figure 6.6: Efficiency plot for unmodified and modified SELFT code with $B_{R,1}(z)$ magnetic field.

in order to depict the influence of the steeper magnetic field profile $B_{R,2}(z)$ with respect to the reduction of dynamic ACI. From Fig. 6.8, it can be seen that the undesired parasitic mode ($TE_{-26,8}$ mode at $\approx$130.21 GHz) appears clearly in the spectrum plot for the non-uniform distribution of the $B_{R,1}(z)$ magnetic field profile. The corresponding zoomed figure of the encircled region of Fig. 6.8 has also been plotted in order to have a clear visualization of the appearance of the parasitic mode $TE_{-26,8}$. On the other hand calculations with the non-uniform distribution $B_{R,2}(z)$ of the magnetic field profile shows that the same parasitic mode i.e. $TE_{-26,8}$ has less influence on dynamic ACI (see Fig. 6.9) but by taking the zoom picture of the encircled region, shown in Fig. 6.9, one can observe that also another parasitic mode $TE_{-28,9}$ is oscillating at $\approx$132.78 GHz due to dynamic ACI. This fact can also been observed in the field profile plot shown in Fig. 6.5. This comparison of ACI for different magnetic field profiles shows that in the design of high-power gyrotron cavities always both the geometry of the uptaper waveguide and the distribution of the magnetic field must be included in order to minimize the ACI. SELFT code calculations shows that the achievable output power in the main operating mode is $\approx$800 kW using above mentioned beam parameters. The parasitic modes have an impact as dynamic ACI within 1% of the power in the main mode ($\approx$7.8 kW).

For the purpose of analyzing the effect of variation of two different magnetic fields with respect to the output power delivered by the main mode $TE_{-28,8}$ as well as by the parasitic TE modes $TE_{-26,8}$ and $TE_{-28,9}$, a comparison has been plotted in Fig. 6.10. Here again only those modes

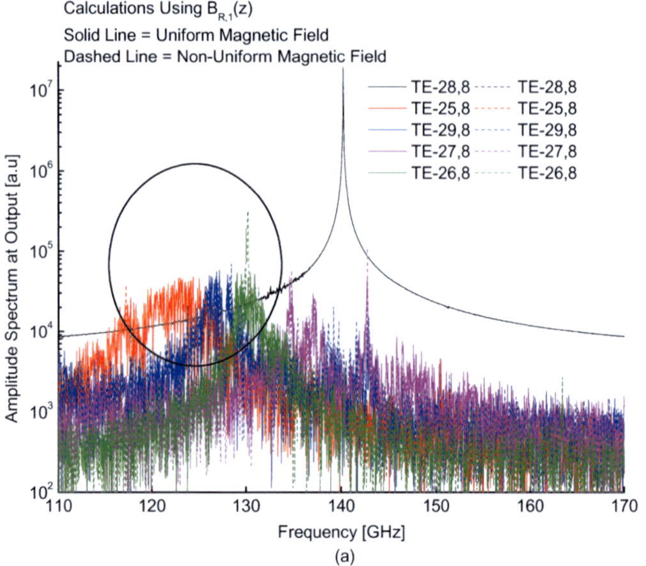

(a)

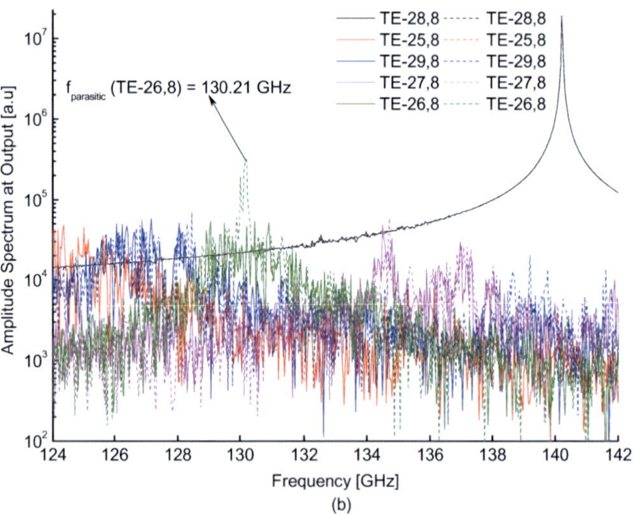

(b)

Figure 6.7: (a) RF spectrum of the modes at the geometry output corresponding to Fig. 6.3 (b) Zoomed picture of the encircled portion of (a).

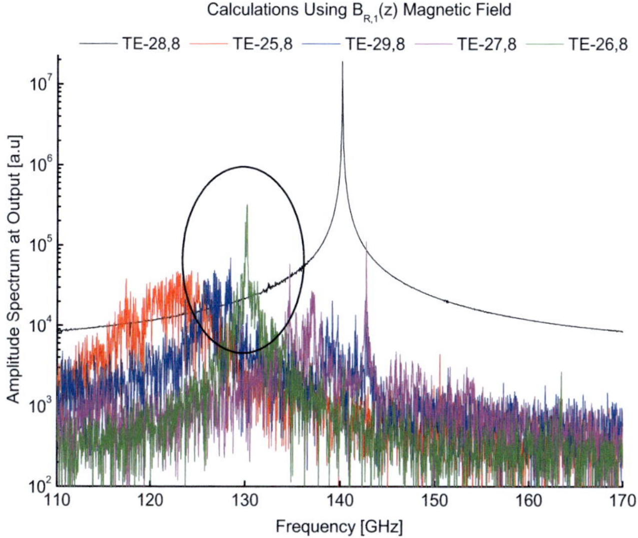

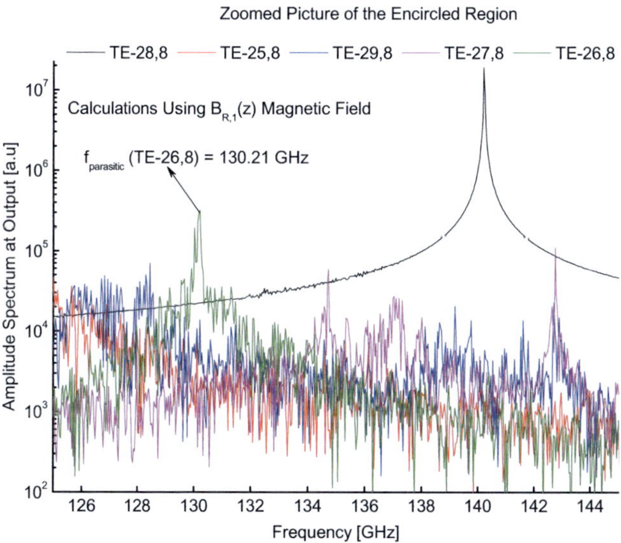

Figure 6.8: RF spectrum of the modes at the geometry output corresponding to Fig. 6.5 calculated using $B_{R,1}(z)$ magnetic field profile. The corresponding zoomed picture of the encircled region is also presented.

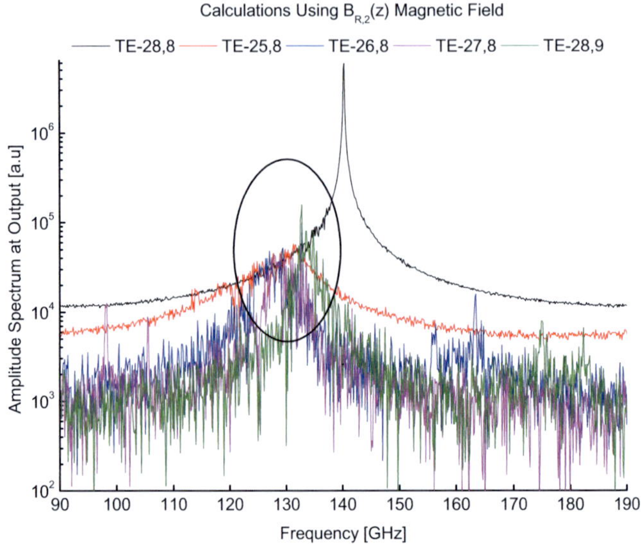

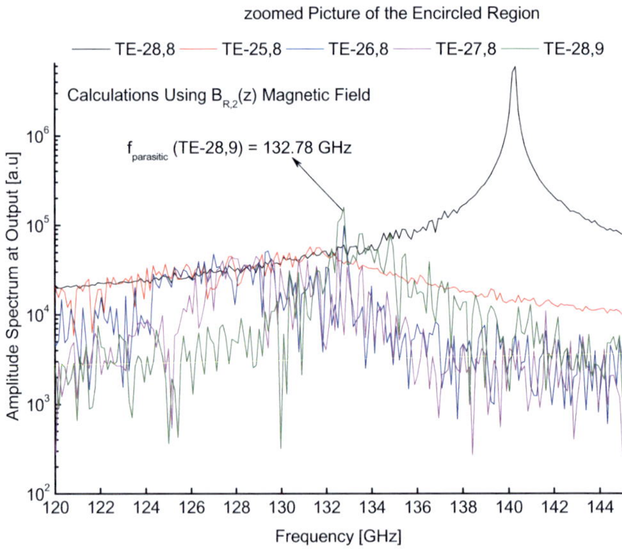

Figure 6.9: RF spectrum of the modes at the geometry output corresponding to Fig. 6.5 calculated using $B_{R,2}(z)$ magnetic field profile. The corresponding zoomed picture of the encircled region is also presented.

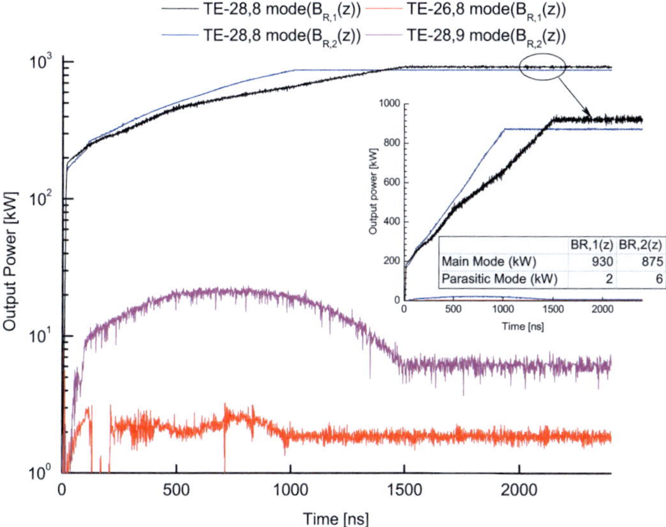

Figure 6.10: Comparison of output power over time after switching on nominal operation parameters for main and parasitic mode using two different magnetic field profiles $B_{R,1}(z)$ & $B_{R,2}(z)$. Zoomed figure of the output power over time, plotted in linear scale, to visualize the power difference in the main modes.

are shown which are useful for comparison because all other modes have output powers close to ≈0.01 kW. A power comparison table between main mode and parasitic mode due to two different magnetic field profiles is also shown in Fig. 6.10. In these calculations, the beam parameters are: $V_{beam} = 81.8$ kV, $I_{beam} = 43.2$ A,$\alpha = 1.18$, $B_0 = 5.615$ T and $R_{beam} = 10.17$ mm. From the zoomed plot of the encircled portion as shown in Fig. 6.10 it can be observed that, for the $B_{R,2}(z)$ magnetic profile, the output power of the main mode has been reduced to ≈875 kW in comparison to the output power delivered by the main mode considering the $B_{R,1}(z)$ magnetic profile field (≈930 kW). Similarly it can also be observed, from the main plot shown in Fig. 6.10, that the output power of the parasitic mode $TE_{-28,9}$ due to the $B_{R,2}(z)$ magnetic field profile has a significant power level as compared to the parasitic mode $TE_{-26,8}$ due to the $B_{R,1}(z)$ magnetic field profile. As expected, the electron efficiency is also affected as shown in Fig. 6.11 being reduced as highlighted in the zoomed figure of the encircled portion of the Fig. 6.11. Just passing-by, we also note the difference in the transients generated by these two magnetic field configurations.

A possible interpretation of the above observations is: Due to the fact that the parasitic mode $TE_{-28,9}$ has the same azimuthal index as the main mode $TE_{-28,8}$, the coupling is strong enough to share energy from the electron beam.

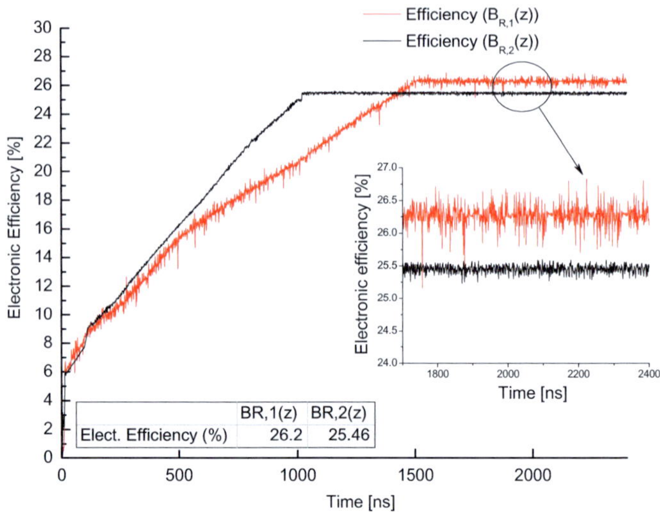

Figure 6.11: Comparison of the efficiency of energy exchange calculated with $B_{R,1}(z)$ & $B_{R,2}(z)$ magnetic field profiles with the same operational parameters as for Fig. 6.10.

All the above calculation are carried out for a particular numerical parameter setting like spatial discretization (0.1 mm), time discretization (0.01 ns), azimuthal discretization (29 phases) etc and also for a particular operating parameters like beam voltage, beam current, beam radius, maximum magnetic field etc. This parametric study with the modified SELFT code has been carried out in order to depict the effect of variations of the other various parameters (numerical as well as physical) on the beam-wave interaction calculations and has been described in chapter 5.

6.1.1 Comparison between the simulated and the experimental results

Using the modified version of the SELFT code with the non-uniform distribution of the magnetic field profile $B_{R,1}(z)$ in the gyrotron interaction calculations as well as with $\Delta t = 0.01$ ns (integration time step), attempts were made to reproduce the measured parasitic spectra [SCG$^+$11] for the following operating point: $V_{beam} = 81.8$ kV, $I_{beam} = 43.2$ A, $\alpha = 1.18$, $B_0 = 5.615$ T and $R_{beam} = 10.17$ mm. In this case 44 TE modes (both co-rotating and counter-rotating) are considered in the simulation. Fig. 6.12 shows the simulated electric field profile, observed at 2300 ns simulation time, of different modes which are relevant for our comparison study. The electric field strength of the other competing modes has been observed significantly lower than the strength of the main operating mode electric field profile and hence may be considered at the noise level. Cavity and uptaper geometry profile along with magnetic field profile for the W7-X 140 GHz gyrotron series SN4R, considered in this study, is shown in Fig. 6.2. The very week oscillating pattern in the electric field profile of the main mode observed in the uptaper region reflects a beating process between two waves of the same mode at different frequencies (see Fig. 6.12). In addition to the main mode oscillation ($\approx$140.2 GHz), some other parasitic modes ($TE_{-28,7}$, $TE_{-28,9}$, $TE_{-26,8}$, $TE_{-28,9}$ and $TE_{-24,9}$) with less electric field strength are also observed in the uptaper section of the geometry (see Fig. 6.12). This indicates the power exchange between RF travelling waves and the spent electron beam in the uptaper section of the geometry. As a result dynamic ACI phenomena are observed. The RF spectrum (see Fig. 6.13), recorded at $z = 77.0$ mm and simulation time 2300 ns, reveals that, in addition to the main operating cavity frequency, some other parasitic frequencies were also observed which were compared with the measured experimental values. These parasitic frequencies $\approx$136.05 GHz, $\approx$133.81 GHz, $\approx$129.66 GHz and $\approx$126.45 GHz correspond to the $TE_{-28,7}$, $TE_{-28,9}$, $TE_{-26,8}$, $TE_{-28,9}$ and $TE_{-24,9}$ modes respectively, shown in the RF spectrum plot, are compared with the experimentally measured values (see Fig. 6.13). Simulations using only 20 neighbour modes of the operating cavity mode gave the same result so that in the following studies 20 TE modes have been considered in the calculations.

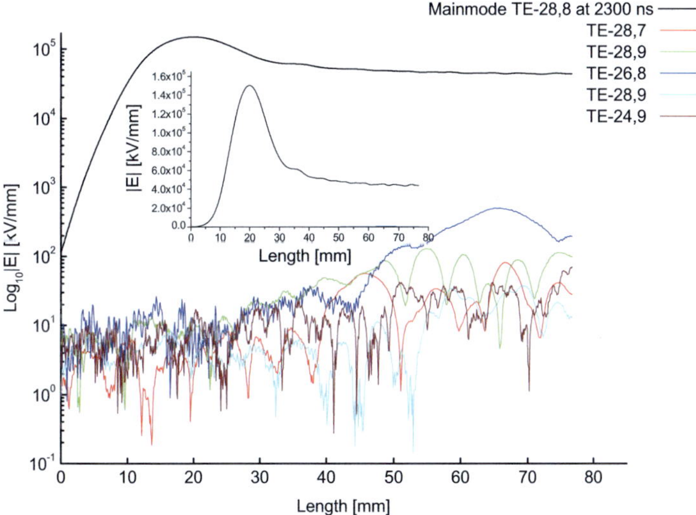

Figure 6.12: Simulated multi-mode electric field profiles obtained using the modified version of the SELFT code for the W7-X Series SN4R gyrotron. Electric field profile is also plotted in a linear scale to visualize the very week oscillating behavior in the main operating mode.

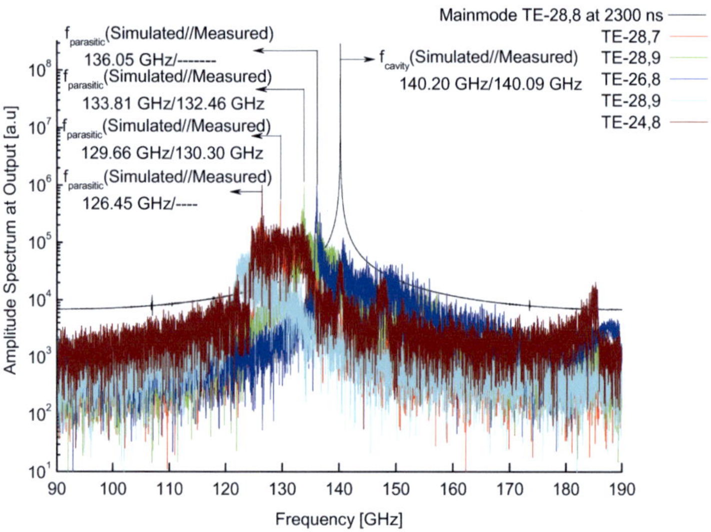

Figure 6.13: RF spectrum at the geometry output for W7-X Series SN4R comparing simulated as well as measured parasitic frequencies.

From Fig. 6.13 it is also observed that the spectrum amplitudes of the parasitic modes are significantly lower than the spectrum amplitude of the main operating mode. These results are very promising and have an excellent agreement between the simulated as well as experimental results [SCG+11]. This confirms that the inclusion of the effect of the non-uniform distribution of the magnetic field in the beam-wave interaction calculations is necessary to describe the dynamic ACI in the uptaper section of gyrotron resonators. According to the modified SELFT simulation result, the achievable output power in the main operating mode measured at the resonator output as well electronic efficiency are $\approx$930 kW and $\approx$32% respectively with $\approx$2 kW ($\approx$0.2%) power being propagating in the parasitic $TE_{-26,8}$ mode (see Fig. 6.10).

6.2 170 GHz 1 MW $TE_{-32,9}$ mode conventional cavity gyrotron

In this section multi-mode self-consistent simulations for the EU 170 GHz 1 MW $TE_{-32,9}$ mode conventional cavity gyrotron for ITER are presented. In these calculations the modified version of the SELFT code, which includes the non-uniform distribution of the magnetic field profile, has been utilized in order to investigate the presence of spurious/parasitic oscillations in the output section of the gyrotron resonator which may be attributed to dynamic ACI phenomena. In this work the optimized conventional cavity geometry as well as the magnetic field profile has been used as input. Fig. 6.14 shows the cavity profile along with the non-linear uptaper section and magnetic field profile which are taken into account into this study.

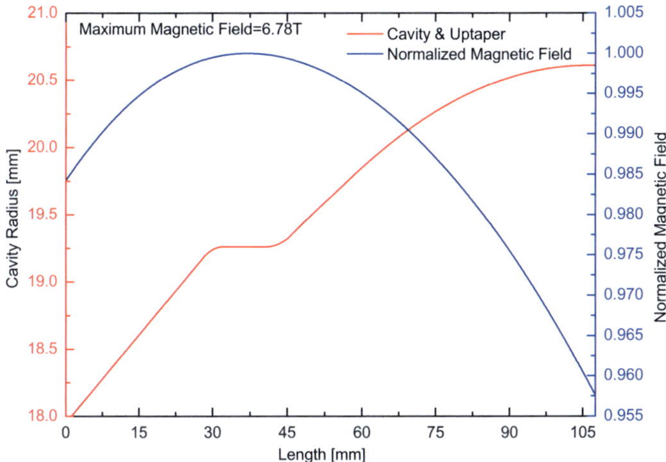

Figure 6.14: Optimized conventional cavity geometry including nonlinear uptaper as well as magnetic field profile for the 170 GHz $TE_{-32,9}$ mode gyrotron.

In order to study possible ACI in the uptaper region the self-consistent multi-mode calculations are performed using the beam parameters $V_{beam} = 79.5$ kV, $I_{beam} = 40.0$ A, $\alpha = 1.30$, $B_0 = 6.785$ T and $R_{beam} = 9.44$ mm. Multi-mode electric field profiles, which are obtained by including up to 20 competing TE modes, calculated up to 3000 ns simulation time using the non-uniform distribution of the magnetic field profile are shown in Fig. 6.15. Here again only those modes are considered which are relevant for our study because all other modes appear with very low magnitudes closer to the noise level, as compared to the considered modes. These electric field profiles are taken at 2500 ns simulation time when the dynamic ACI effect is most evident.

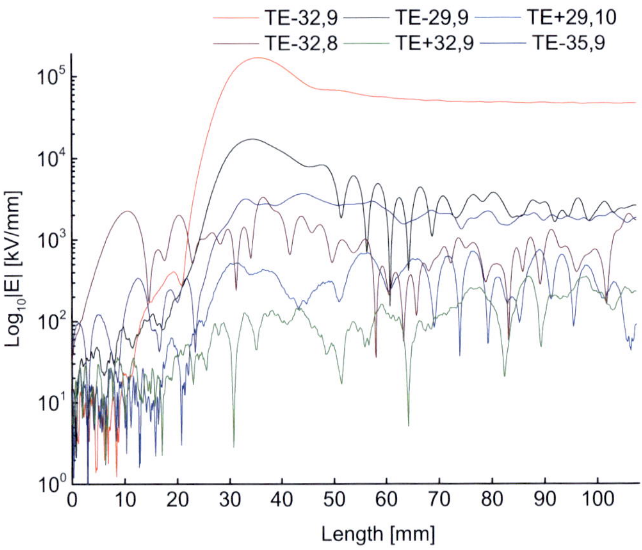

Figure 6.15: Multi-mode electric field profiles obtained using the modified version of the SELFT code with a varying magnetic field profile.

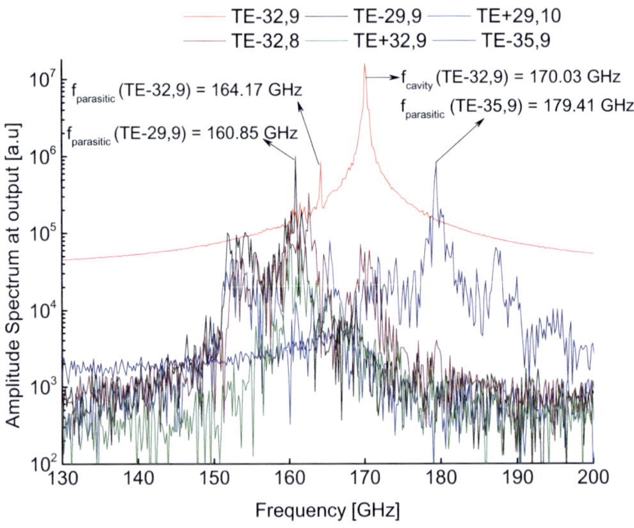

Figure 6.16: RF spectrum of the relevant modes at the geometry output corresponding to Fig. 6.15.

The simulated electric field profiles, shown in Fig. 6.15, depict that, in addition to the required operating mode $TE_{-32,9}$, the parasitic modes $TE_{-29,9}$ and $TE_{-35,9}$ also oscillating in the cavity and output geometry while on the contrary, this does not happen if we consider a uniform magnetic field profile.
In order to make a clear confirmation of the appearance of the above parasitic modes in the uptaper section of the geometry, the RF spectrum of the simulated electric field profiles is shown in Fig. 6.16, observed at 2500 ns simulation time and at the axial coordinate $z = 99.0$ mm. From Fig. 6.16 one can confirm that in addition to the desired oscillating mode $TE_{-32,9}$, oscillating at $\approx$170.03 GHz, also the parasitic modes $TE_{-29,9}$ and $TE_{-35,9}$ appear at $\approx$160.85 GHz and $\approx$179.41 GHz, respectively. The RF spectrum plot also shows the appearance of the low amplitude parasitic mode $TE_{-32,9}$ oscillating at $\approx$164.17 GHz which leads to the presence of a beating process with the main operating mode. This fact is also reflected by a weak modulating feature in the electric field profile of the main mode (see Fig. 6.15). Thus according to the simulation results minor dynamic ACI phenomena have been observed in the 170 GHz $TE_{-32,9}$ 1 MW gyrotron while experimental results are still to be performed in order to have benchmarking of the above simulated results.
Finally it is required to mention that with the modified version of the SELFT code the achievable output power for the parameter set utilized here is calculated to be $\approx$1.02 MW in the selected operating mode $TE_{-32,9}$. Parasitic modes have an impact within approximately 0.6% of the main operating mode power. The efficiency calculated by multi-mode simulations drops in comparison to single-mode simulations from 38% to 35%, because the upper region of the stability area (accelerating voltage over magnetic field plane) can not be reached.

6.3 170 GHz 2 MW $TE_{-34,19}$ mode coaxial cavity gyrotron

In this section multi-mode self-consistent calculations using the modified version of the SELFT code are described with respect to investigate the existence of dynamic ACI in the output section of the 170 GHz 2 MW $TE_{-34.19}$ mode coaxial cavity gyrotron resonator. In these calculations, the beam parameters are: $V_{beam} = 89.0$ kV, $I_{beam} = 78.0$ A, $\alpha = 1.20$, $B_0 = 6.82$ T and $R_{beam} = 10.2$ mm. These beam parameters are chosen based on the experimental values. The coaxial cavity geometry as well as the magnetic field profile considered in this simulation study are shown in Fig. 6.17. Initially the gyrotron cavity was optimized and designed by IHM/KIT under the assumption of a constant magnetic field in the unmodified SELFT code.
In order to investigate ACI phenomena, a multi-mode simulation up to 2400 ns simulation time was performed including all the proposed modifications implemented in the modified KIT SELFT code. Multi-mode electric field profiles considering up to 20 competing TE modes calculated using the non-uniform distribution of the magnetic field are shown in Fig. 6.18. Here only the relevant modes are shown; all other modes have very low magnitudes, closer to the noise level, as compared to the TE modes shown in Fig. 6.18.

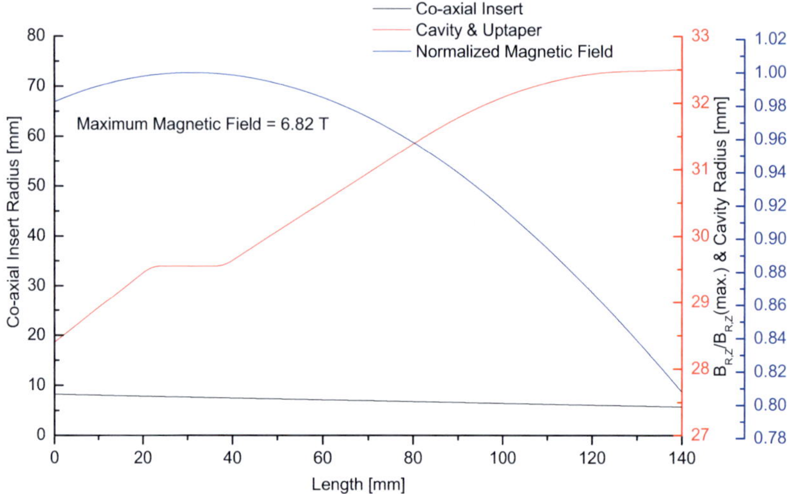

Figure 6.17: Optimized coaxial cavity geometry including nonlinear uptaper as well as magnetic field profile for the 170 GHz $TE_{-34,19}$ mode gyrotron.

The calculated electric field profiles shown in Fig. 6.18 depict that practically no other modes than the required operating mode $TE_{-34,19}$ are oscillating as spurious/parasitic modes in the cavity uptaper section with a significant power level. With a more closer look one can observe that the $TE_{+31,20}$ mode is also oscillating in the uptaper section but having a very low magnitude of the electric field. The relative output power in the $TE_{+31,20}$ mode only is 0.12%. As a result, this mode cannot be considered to have a significant effect as dynamic ACI phenomenon. While on the contrary, this does not happen if we consider a uniform magnetic field profile in the beam-wave interaction calculations. This reveals that the weak appearance of the $TE_{+31,20}$ mode is due to the consideration of the non-uniform magnetic field profile. These theoretical results were also been confirmed experimentally by measurements at IHM/KIT. No spurious/parasitic oscillations with a significant impact which may be due to dynamic ACI [PDD$^+$04, PDD$^+$05] were observed. This confirms that the simulated results with the modified version of the SELFT code, presented in this section, have a good agreement with the experimental results. In order to have a more closer look about the presence of any oscillating mode behaving as a dynamic ACI in the uptaper section of the geometry, the RF spectrum of the simulated electric field profiles (see Fig. 6.18) using the non-uniform distribution of the magnetic field profile in the modified version of the SELFT code is presented in Fig. 6.19. The RF spectrum plot is recorded at $z = 100.0$ mm as well as at 1875 ns simulation time. In this

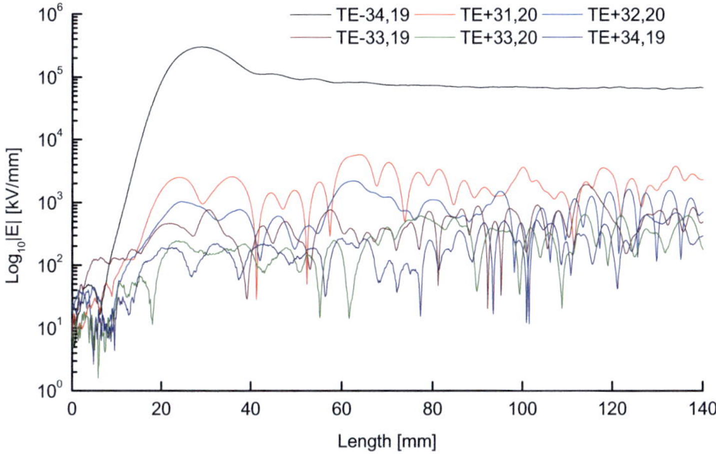

Figure 6.18: Multi-mode electric field profiles obtained using the modified SELFT code with a non-uniform magnetic field profile.

case again only those modes are shown which are relevant for our study. From this RF spectrum plot it can also be seen that no other modes than the required cavity mode, oscillating at 170 GHz, appear as spurious/parasitic modes with significant amplitude level so that they can be attributed to dynamic ACI. The mode $TE_{+31,20}$ oscillating at 164 GHz has been found with a very low intensity which also corresponds to the simulated electric field profile. Thus from the above discussion it can be concluded that both the geometry as well as the magnetic field profiles, used in this study, are well suited to have no spurious/parasitic oscillations. The same facts are also reflected in the experimental results which have been performed at IHM/KIT.

Finally it is required to mention that the simulated result with the modified version of the SELFT code considering up to 20 competing TE modes and with the above mentioned beam parameters shows the achievable output power of $\approx$2.2 MW in the selected operating mode $TE_{-34,19}$ which is in agreement with the experimental findings. Parasitic modes have impact as a dynamic ACI with approximately 0.15% of the output power in the main operating mode.

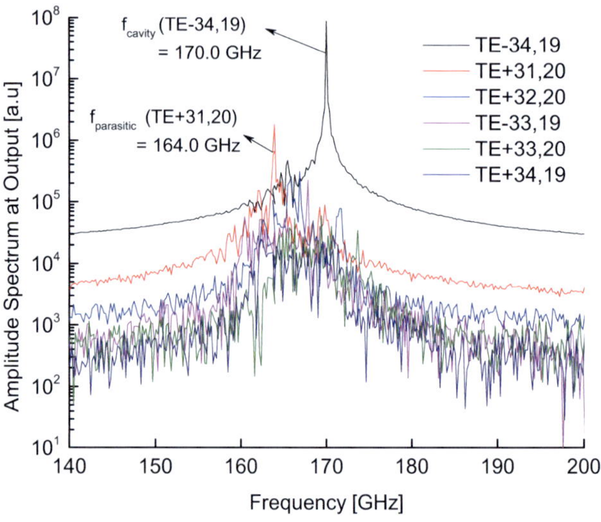

Figure 6.19: RF spectrum of the relevant modes at the geometry output corresponding to Fig. 6.18.

6.4 240 GHz 2.5 MW TE$_{-55,29}$ mode coaxial cavity gyrotron

A first step towards a sample cavity design for a very high frequency, very high order mode 240 GHz high-power gyrotron has been performed in order to investigate the influence of the non-uniform distribution of the magnetic field in terms of the presence of the dynamic ACI effect. A coaxial cavity is used for the beam-wave interaction. The main restriction is the ohmic wall loading on the surface of the coaxial cavity achievable by available cooling and material techniques: The wall loading cannot exceed more than 2.5 kW/cm^2. In order to overcome this problem the size of the resonator is increased and therefore very high order modes make this operation possible. Fig. 6.20 shows the coaxial cavity geometry profile along with the magnetic field profile which are used in this study. For gyrotrons operating with very high-order volume modes, the dense mode spectrum at high eigenvalues delivers a vast number of possible mode candidates and competitors. For the design of a gyrotron with a specific frequency (240 GHz) and output power ($\approx$2.0 MW) several physical and technical requirements have to be met (see Table 1.1). In this numerical experiment a magnetic field profile identical to that of the KIT-Oxford-Instrument-Magnet has been used and in shown in Fig. 6.20.

For the given beam parameters (V_{beam} = 78.5 kV, I_{beam} = 89.0 A, α = 1.20, B_0 = 9.57 T and R_{beam} = 11.4 mm) the output power has been determined using the program called 'MAXPO'[Ker96] which was developed based on the theory of normalized variables [DT86] to

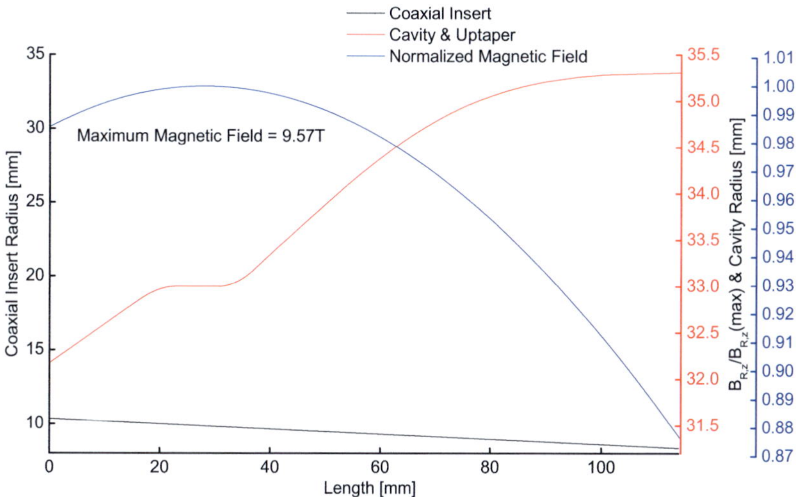

Figure 6.20: Optimized coaxial cavity geometry including nonlinear uptaper as well as magnetic field profiles for an example of a 240 GHz $TE_{-55,29}$ mode gyrotron.

calculate the maximum possible gyrotron output power considering the physical requirements and constraints [KBT04, Ker96]. The possible highest output power over artificial continuous Bessel roots $\chi_{m,p}$ for different normalized resonator lengths μ (range from 16 to 25) with respect to the constraints has been determined keeping in mind the physical requirements and constraints. Within these estimations, no mode conversion has been considered. The highest output power is predicted for $\chi_{m,p}$ values between 160 and 230. Within this range a high number of possible mode candidates are available. Out of those available modes, the $TE_{-55,29}$ mode ($\chi_{m,p} = 165.94$) has been considered as an operation mode for a 240 GHz gyrotron as this mode delivers a suitable power and maintains all those design constraints. The parameters for an optimized sample cavity for the $TE_{-55,29}$ mode with linear input and output tapers together with the length of the parabolic smoothing, can be found in Table 6.1. With the chosen radius of the cavity equal to 33.0 mm an output frequency of 240.20 GHz, slightly higher than the desired 240 GHz has been obtained.

In this work the outer cavity wall is assumed to be smooth without any corrugation for further mode selection. Non-linear tapers have been introduced to fit the cavity geometry to the beam tunnel and launcher section (see Fig. 6.20). These nonlinear tapers have to guarantee low power conversion towards the gun and beam tunnel region as well as lowest possible mode conversion towards the launcher. The synthesis of the non-linear tapers has been performed using a

Parameters	**Values**
$L_d/L_R/L_u$ (mm)	20 / 14 / 10
$\theta_d/\theta_R/\theta_u$ (°)	2.4 / 0 / 3.2
Smoothing $L_{s,d}/L_{s,u}$ (mm)	3.2 / 3.2
$R_R/R_i/R_b$ (mm)	33 / 10 / 11.4
θ_i (°)	-1

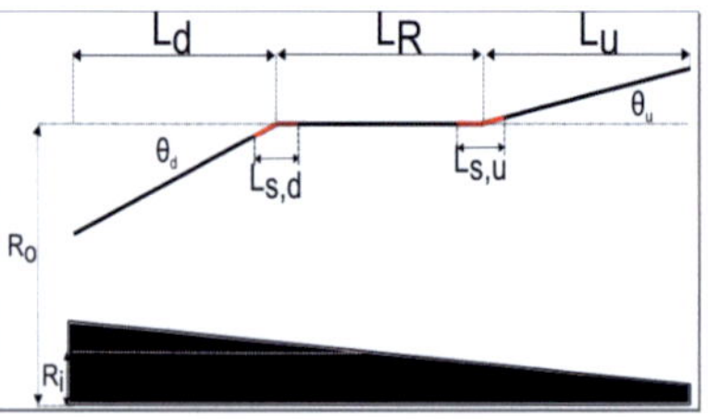

Table 6.1: Optimized sample cavity geometry for the $TE_{-55,29}$ mode (for parameters see schematic diagram shown in the right hand side of this table).

	Position 1	Position 2	Position 3
Reflection $TE_{-55,29}$	99.71	06.47	99.49
Reflection $TE_{-55,28}$	00.27	00.01	00.45
Reflection $TE_{-55,30}$	-	-	0.017
Reflection $TE_{-55,29}$	-	93.38	-
Reflection $TE_{-55,28}$	0.007	0.08	-
Reflection $TE_{-55,30}$	-	0.011	-
Position 1: The junction between downtaper and straight portion of the cavity. Position 2: The junction between uptaper and straight portion of the cavity. Position 3: The output of the cavity.			

Table 6.2: Mode conversion [%] at the different positions for the optimized nonlinear tapers

scattering matrix program based on eigenwave expansion (NTLAP-based on the generalized scattering matrix method). The optimized geometry shows low power conversion (less than 0.1%) towards the gun region. The non-linear uptaper starting from the end of the cylindrical cavity (up to 42.0 mm) has a length of 73.0 mm. The total length of the cavity including all tapers is 115.0 mm, which is comparable with the designs for the existing IHM/KIT European 170 GHz 2 MW (≈120 mm) gyrotron for ITER. The corresponding reflection coefficients values can be found in Table 6.2.

In the next step, time-dependent start-up simulations are performed considering a linear rise of the accelerating voltage. In order to investigate the influence of the non-uniform distribution of the magnetic field in terms of ACI phenomena in the designed 240 GHz cavity, self-consistent multi-mode simulations (with 20 modes) are performed using two different versions of the SELFT code i.e. SELFT with uniform magnetic field and SELFT with non-uniform magnetic field. Comparison of the electric field profiles are shown in Fig. 6.21. Here only those modes are shown which are retained relevant to our comparisons. The electric field profiles in case of a uniform magnetic field profile (shown by solid line) which corresponds to the maximum value of the magnetic field profile as well as in the case of a non-uniform magnetic field profile

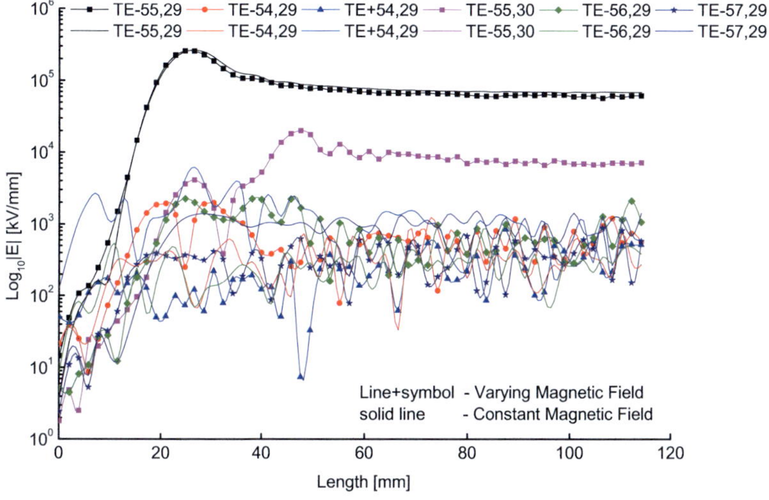

Figure 6.21: Comparison between field profiles obtained in multi-mode calculation using the modified SELFT code with uniform and non-uniform magnetic fields profile.

(shown by line + symbol) recorded at simulation time 2300 ns are shown in Fig. 6.21. The figure shows that, due to the consideration of a non-uniform magnetic field profile in the beam-wave calculations, in addition to the operating mode $TE_{-55,29}$, an undesired parasitic mode $TE_{-55,30}$ is also oscillating in the cavity uptaper region, with a significant power level. While on the other hand it is not observed in the case of a uniform magnetic field (see Fig. 6.21). This undesired oscillations can be attributed to the fact that the inclusion of the non-uniform magnetic field in the SELFT calculations which has significant different values from the magnetic field value at the mid of the cavity shows the existence of dynamic after cavity interaction phenomenon. One can also observe from Fig. 6.21 that in this study the electric field profiles of the operating mode, for both the cases, are nearly identical.

In order to confirm the above mentioned undesired oscillations, the RF spectrum, corresponding to the electric field profiles (see Fig. 6.21), have been plotted shown in Fig. 6.22. In this study, RF spectra, recorded at $z = 110.0$ mm, are obtained using the uniform magnetic field corresponding to the maximum value (solid line) as well as the non-uniform magnetic field (line + symbol) profile. It should be mentioned that the spectrum were observed at the simulation time 2300 ns. In Fig. 6.22, only those modes are shown which remain relevant for our comparison study. From Fig. 6.22, one can observe that, due to the consideration of the non-uniform magnetic field profile, in addition to the desired operating mode $TE_{-55,29}$

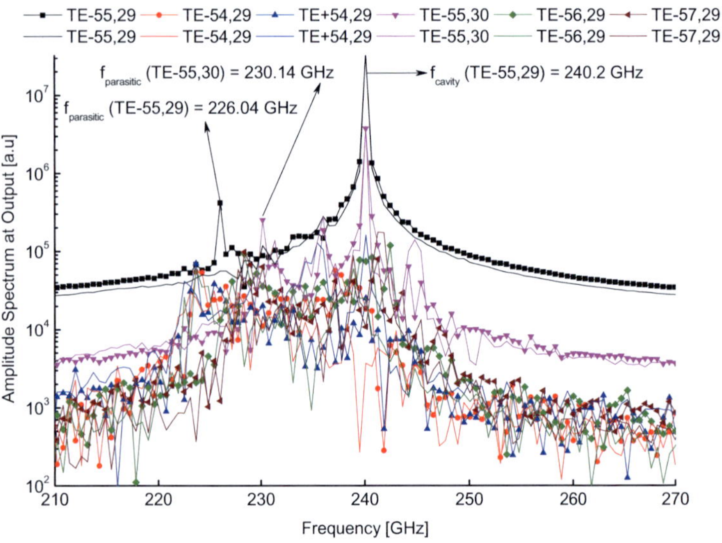

Figure 6.22: RF spectrum of the relevant modes at the geometry output corresponding to Fig. 6.21.

mode, oscillating at ≈240.2 GHz, also another parasitic mode is oscillating at ≈230.14 GHz ($TE_{-55,30}$ mode) which has a significant impact as a dynamic ACI phenomenon. This fact is also revealed in the electric field profiles (see Fig. 6.21). The RF spectrum plot also shows the appearance of the parasitic mode $TE_{-55,29}$ oscillating at ≈226.04 GHz corresponding to the presence of a beating process with the desired operating mode. This fact is also reflected by the presence of the modulating feature in the electric field profile of the desired mode calculated with the non-uniform magnetic field distribution (see Fig. 6.21). These undesired ACI Phenomena are attributed to the consideration of the influence of a non-uniform distribution of the magnetic field in the differential equations (2.8 and 2.29) necessary for the gyrotron beam-wave interaction calculations. Whereas, on the other hand, calculations with a uniform magnetic field does not show the appearance of any undesired parasitic mode which may has any significant effect as an ACI phenomena. These calculations also confirm that the detection of the presence of ACI phenomena in the uptaper region of the gyrotron cavity in the simulation result is due to the consideration of the non-uniform magnetic field explicitly in the self-consistent beam-wave interaction calculations. These simulation results again reinforce the fact that in order to determine the occurrence of undesired parasitic oscillations which can be seen as dynamic ACI phenomena in the output section of the resonator, it is necessary to consider the influence of the non-uniform distribution of the magnetic field in the beam-wave interaction calculations.

Finally it should be mentioned that the initial optimization with the modified SELFT code shows the achievable output power of $\approx$2.1 MW (for the optimistic upper parameter limits in Table 1.1) with the selected operating mode $TE_{-55,29}$. In this case parasitic modes have an impact as a dynamic ACI within $\approx$1.2% of the output power in the main operating mode. The efficiency drops in comparison to single mode simulations from 38% to 33%, because the upper region of the stability area (accelerating voltage over magnetic field plane) can not be reached due to mode competition. This is because due to the presence of the spurious oscillations which slightly lower the achievable interaction capability of the main mode, but still a satisfying value for the efficiency of 33% is attainable. Final mode selection and optimization of the 240 GHz gyrotron cavitiy with many more competing modes and with different working operating points is still required.

7 Conclusions and outlook

In this work, dynamic ACI phenomena in the uptaper region of high power gyrotron cavities (conventional as well as coaxial) have been investigated. Goal of this investigation was the study of the influence of the non-uniform distribution of the magnetic field on the parasitic/spurious oscillations present in the output section of gyrotron resonators which are attributed as dynamic ACI phenomena.

First an adiabatic approximation has been used in order to depict the region(s) of fulfilment of the condition for the electron beam-wave interaction by solving the well known electron cyclotron resonance equation. The solution includes varying electron velocity $u_{\parallel}(z)$ according to the non-uniform distribution of the magnetic field. Within the validity of this approach, numerical experiments have been performed with two different KIT gyrotron configurations i.e. the W7-X 140 GHz $\mathrm{TE}_{-28,8}$ mode gyrotron as well as the $\mathrm{TE}_{-22,8}$ mode 140 GHz step-frequency tunable gyrotron. In both the cases, region(s) in the uptaper section of the cavity geometry where found as possible regions for the appearance of dynamic ACI phenomena due to the fulfilment of the resonance condition. In the adiabatic constant implementation, only the electron transverse momentum equation has been solved for the beam-wave interaction calculations. Dynamic ACI phenomena with frequencies lower than that of the desired main mode oscillating frequency have been observed in RF spectral plots. In addition to the above, interacting frequency curves are also estimated for 900 macro-particle and 13 $\mathrm{TE}_{m,p}$ modes, having different γ and $\beta_{\parallel}(z)$ for each macro-particle (see chapter 3). From this numerical experiment it was observed that the upper high-frequency gyro-TWT branches are broadened due to strong sensitivity to different (Doppler broadening) and they are much less prone to dynamic ACI.

As main new contribution the present study includes the enhancement of the capability of the physical interaction models implemented in simulating tools like the time-dependent, self-consistent KIT SELFT multi-mode code in order to describe the dynamic ACI phenomena as close as possible. It includes the implementation of the modifications in the differential equation for the electron momentum as well as for the RF field profile (see chapter 4). These modifications mainly include the introduction of the term $dB_{R,z}/dz$ in the electron momentum differential equation (solved by Predictor-Corrector numerical method) as well as in the acceleration term of the RF field profile differential equation (solved by Crank-Nicolson numerical method). The modifications also include the implementation of the slowly varying average reference frequency with reference to the axial coordinate of the resonator in order to avoid overflow in the simulations. In addition to that the transit time of electrons has also been

modified which include $v_{\parallel}$ as an average value taken over a number of macro-particles at each discrete point of the cavity geometry. Due to all the implemented modifications (see chapter 4), the computational time has increased by $\approx$40% as compared to the unmodified version of the SELFT code reported in [Ker96]. The main contribution is due to the newly introduced non-uniform magnetic field distribution particularly in the electron momentum equation.
Convergence studies have been preformed with respect to three numerical parameters such as spatial discretization step size, temporal discretization step size as well as number of electrons require for initial azimuthal phase discretization. These studies showed the convergence of the solutions.
The new capability of the modified KIT SELFT code has been applied for dynamic ACI studies using four different gyrotron configurations (two conventional cavity gyrotrons and two coaxial cavity gyrotrons). These studies confirmed that an undesired interaction in the uptaper region can result in additional parasitic oscillations with relative power lies within the 1% region. It is also concluded that due to implementation of $dB_{R,z}/dz$, some spurious oscillations have been observed in the RF spectral plot due to significant variation of the magnetic field towards the geometry output. Initial comparison with experiments has been performed for verification of the modified SELFT code. Attempts were made to reproduce the measured parasitic spectra [SCG^{+}11] for the $TE_{-28,8}$ mode 140 GHz W7-X gyrotron developed for the W7-X Stellarator. In this case theory and experiment are in good agreement. This agreement opens the possibility to simulate the physical behaviour of the gyrotron beam-wave interaction phenomenon more closely in the uptaper section of the geometry in the presence of the non-uniform distribution of the magnetic field. Investigations of dynamic ACI phenomena with the modified version of the KIT SELFT code, which includes axial inhomogeneous magnetic fields distribution, are made possible for the first time. In addition to the above, it is also concluded that the radius profile of the uptaper section of the geometry is very important in avoiding the presence of the dynamic ACI phenomena. As a result, it has been observed that the shape of the uptaper section profile used in the 170 GHz $TE_{-34,19}$ mode coaxial cavity gyrotron as well as in the 170 GHz $TE_{-32,9}$ mode conventional cavity gyrotron is less prone to ACI parasitic/spurious oscillations than the uptaper of the 140 GHz $TE_{-28,8}$ mode W7-X gyrotron.
In addition to that a preliminary study has been performed to investigate the influence of $dB_{R,z}/dz \neq 0$ in terms of the presence of dynamic ACI effect in the uptaper section of very high frequency (for example 240 GHz) and very high order mode (for example $TE_{-55,29}$) gyrotron. The RF spectrum plot shows that in addition to the operating mode $TE_{-55,29}$, an undesired parasitic mode $TE_{-55,30}$ is also oscillating in the cavity uptaper region, with a significant power level ($\approx$1.2% of the main mode power). On the contrary this does not happen if we consider constant magnetic field. These undesired oscillations can be attributed to the consideration of the non-uniform magnetic field in the SELFT multi-mode calculations.

These studies suggest that the occurrence of dynamic ACI is caused due to the changing magnetic field towards the geometry output where it is significantly different from the value at the centre of the cavity.

During the next development phase, on the theoretical side, several extensions of the mathematical model have been reviewed (see chapter 2) that have not been included in the modified version of the KIT SELFT code. These are the consideration of mode conversion and the inclusion of TM modes (e.g. for relativistic gyrotrons) which includes a full description of the RF field profile. It is also necessary to make further improvement in the mathematical modelling of the electron motion equation both in transverse as well as axial directions. This includes consideration of the imaginary part of $\vec{u}_\perp$ (see equation (2.32)) also in the calculations of the electron motion equation as well as in the excitation term ($\vec{R}_k$) of the RF field equation (see equation 2.33b).The underlying assumption in SELFT of using slowly varying RF field profiles is questionable for the ACI envelope, since a relative fast beating is present.
To address this point and also for a more precise comparison with experimental results, it is necessary to abandon some of the simplifications of the set of equations in modified SELFT code (i.e. the way we model the electron movement which includes replacement of the time coordinate through the space coordinate by using equation (2.28)), in favour of, a filled-cavity quasi-Particle In Cell (PIC) like approach similar to the approach adapted in EURIDICE quasi PIC (filled cavity approach) as well as in TUHH code [Jel00]. In this approach in the modified SELFT code it is necessary to implement directly the numerical solution of the Maxwell equations as well as Lorentz force equation in the two-dimensional phase-space. The obtained RF field values can directly be interpolated on to the simulation particles which will be advanced by a second order Boris scheme. From the advanced particle position, charge and current source terms can then be interpolated onto the mesh points in order to compute the RF field profile values at each mesh points of the simulation domain. The calculated RF field profile values are then interpolated onto the particle positions. The above approach is required to be performed at each integration time step. This approach can easily be implemented with the existing utilized mathematical formalism in the SELFT code to perform much faster simulations than with Particle-In-Cell (PIC) codes. It was mentioned that in the SELFT code the RF field profile at each axial coordinate value is known prior the movement of the electrons throughout the simulation domain whereas these quasi-PIC like approach field solvers require no prior knowledge about the electric fields. In this way we can abolish the 'fixed-field'approach to make the simulations more accurate.

List of Figures

List of Tables

Bibliography

[ABS+08] F. Albajar, T. Bonicelli, G. Saibene, S. Alberti, D. Fasel, T. Goodman, J.P. Hogge, I. Pagonakis, L. Porte, M. Q. Tran, K. A. Avramides, J. Vomvoridis, R. Claesen, M. Santinelli, O. Dumbrajs, G. Gantenbein, S. Kern, S. Illy, J. Jin,B. Piosczyk, T. Rzesnicki, M. Thumm, M. Henderson, S. Cirant, G. Latsas, And I. Tigelis. Review of the European programme for the development of the gyrotron for ITER. *Proceedings of the 15th Joint Workshop (Electron Cyclotron Emission And Electron Cyclotron Resonance Heating (Ec-15)*, Yosemite National Park, California, USA, pp-415-421, 10-13 March 2008.

[AIP+08] K. A. Avramides, C. T. Iatrou, I. G. Pagonakis, and J. L. Vomvoridis. *EURIDICE.* National Technical University, Athens, unpublished, 2008.

[AS64] Milton Abramowitz and Irene A. Stegun. *Handbook of Mathematical Functions with Formulas, Graphs, and Mathematical Tables. Chapter 9 and 10*, Dover, New York, ninth dover printing, tenth GPO printing edition, 1964.

[ATA+11] S. Alberti, T. M. Tran, K. A. Avramidis, F. Li, and J.-P. Hogge. Gyrotron parasitic-effects studies using the time-dependent self-consistent monomode code TWANG. *36th International Conference on Infrared, Millimeter, and Terahertz Waves (IRMMW-THz 2011)*, Houston TX USA, October 2, 2011, Tu5.15.

[Bas96] B. N. Basu. *Electromagnetic Theory and Applications in Beam-Wave Electronics*, World Scientific, Singapore, 1996.

[BD04] A. A. Bogdashov and G.G. Denisov. Asymptotic Theory of High-Efficiency Converters of Higher-Order Waveguide Modes into Eigenwaves of Open Mirror Lines. *Radiophysics and Quantum Electronics*, 47(4), pp-283-296, 2004.

[Ber11] Matthias Hermann Beringer. *Design studies towards a 4 MW 170 GHz coaxial-cavity gyrotron.* PhD thesis, Karlsruhe Institute of Technology, Karlsruhe, KIT Scientific Publishing, ISBN 978-3-86644-663-2, 2011.

[BGI+09] Vladimir Bratman, Mikhail Glyavin, Toshitaka Idehara, Yuri Kalynov, Alexey Luchinin, Vladimir Manuilov, Seitaro Mitsudo, Isamu Ogawa, Teruo Saito, Yoshinori Tatematsu and Vladimir Zapevalov. Review of Subterahertz and Terahertz Gyro-devices at IAP RAS and FIR FU. *IEEE Transactions on Plasma Science*, Vol. 37, No. 1, 36-43, January 2009.

[BJ87] E. Borie and B. Jödicke. Self consistent multimode theory for gyrotrons. *Unpublished KFK-primary report* 03.04.02.P09I. 1987.

[BJ90] E.Borie and B.Jodicke. Rieke Diagrams for Gyrotrons. *International Journal of infrared and Millimeter waves*, vol. 11, no.2, 1990.

[BJ92] E Borie and B. Jödicke. Self-consistent theory of mode competition for gyrotrons. *International Journal of Electronics*, 72, pp-721-744, 1992.

[BMP+73] V.L. Bratman, M.A. Moiseev, M.I. Petelin and R.*É*.*É*rm. Theory of gyrotrons with a non fixed structure of the high-frequency field. *Radio physics and Quantum Electronics*, 16, pp-474-480, 1973.

[Bor91] E Borie. *Review of Gyrotron Theory*, KFK-report 4898, 1991.

[Bor93] E.Borie. *Computations of radio-frequency behaviour*, in Edgcombe, C.J. (Hrsg.), Gyrotron Oscillators - Their Principles and Practice, chapter 3, London: Taylor and Francis, 45-86, 1993.

[Bor01] Edith Borie. Effect of Reflection on Gyrotron Operation. *IEEE Transactions on Microwave Theory and Techniques*, vol.49, no.7, pp-1342-1345, July 2001.

[BW86] E. Borie,and H. Wenzelburger. Wall losses in weakly tapered gyrotron resonators. *unveröffentlichter KFK-Primärbericht* 03.04.02.P07B, 1986.

[CAS92] S.Y. Cai, T.M. Antonsen, G. Saraph, and B, Levush. Multi frequency theory of high power gyrotron oscillators. *International Journal of Electronics*, 72, pp-759-777, 1992.

[CB93] R. A. Correa and J. J. Barroso. Space Charge Effects of Gyrotron Electron Beams in Coaxial Cavities. *International Journal of Electronics*, 74, pp-131-136, 1993.

[Che95] F.F. Chen. *Introduction to Plasma Physics*. Springer, New York, USA, 1995.

[CSS+07] E. M. Choi, M. A. Shapiro, J. R. Sirigiri, and R. J. Temkin. Experimental observation of the effect of after cavity interaction in a depressed collector gyrotron oscillator. *Physics of Plasmas*, 14(9), 093302-093305, 2007.

[CSA+11] A.Roy Choudhury, Andreas Schlaich, D. D'Andrea, Stefan Kern, and Manfred Thumm. Influence of Non-Uniform Magnetic Field Distribution on Gyrotron Spurious Oscillations. *36th International Conference on Infrared, Millimeter, and Terahertz Waves (IRMMW-THz 2011)*, Houston TX USA, October 2-7, 2011, Tu5.17.

[CKA+12] A Roy Choudhury, S.Kern, D.D'Andrea and M.Thumm. Numerical Investigations of Parasitic Osicllations in Gyrotron. *39th IEEE International Conference on Plasma Science (ICOPS)*, Edinburgh, Scotland, July 8-12, 2012, 1P-8.

[DK81] A.T. Drobot and K. Kim. Space charge effects on the equilibrium of guided electron flow with gyromotion. *International Journal of Electronics*, 51, pp-351-367, 1981.

[DLM+08] G.G. Denisov, A.G. Litvak, V.E. Myasnikov, E.M. Tai, and V.E. Zapevalov. Development in Russia of High-Power Gyrotrons for Fusion. *Nuclear Fusion*, 48 (5), pp-55-63, 2008.

[DN92] O. Dumbrajs and G.S. Nusinovich. Cold cavity and self-consistent approaches in the theory of mode competition in gyrotrons. *IEEE Transaction on Plasma Science*, Vol. 20, pp.126-132,1992.

[DT86] B. G. Danly and R. J. Temkin. Generalized nonlinear harmonic gyrotron theory. *Physics of Fluids*, 29(2), pp-561-567, 1986.

[Dum01] O. Dumbrajs. *COAXIAL*, Helsinki University of Technology, unpublished, 2001.

[DL92] O. Dumbrajs, and S. Liu. Kinetic theory of electron cyclotron resonance masers with asymmetry of the electron beam in the cavity. *IEEE Transactions on Plasma Science*, 20, pp-133-138. 1992.

[EMR93] G. Engeln-Müllges and F. Reutter. *Numerik-Algorithmen mit ANSI C Programmen*. Mannheim, Scientific-Publishing, Germany, 1993.

[FDJ+99] K. L. Felch, B. G. Danly, H. R. Jory, K. E. Kreischer, W. Lawson, B. Levush, and R..J.Temkin. Characteristics and applications of fast-wave gyrodevices. *Proceedings of the IEEE*, 87(5), pp 752 781, 1999.

[FGP+77] V.A. Flyagin, A.V. Gaponov, M.I. Petelin, and V.K. Yulpatov. The gyrotron. *IEEE Transactions. Microwave Theory and Techniques*, 25, pp-514-521, 1977.

[Fla12] Jens H. Flamm. *Diffraction and Scattering in Launchers of Quasi-Optical Mode Converters for Gyrotrons*. PhD thesis, Karlsruhe Institute of Technology, Karlsruhe, KIT Scientific Publishing, ISBN 978-3-86644-822-3, 2012.

[FN88] V.A. Flyagin, and G.S. Nusinovich. Gyrotron oscillators. *Proceedings of the IEEE*. 76, pp-644-656, 1988.

[FR81] A.W. Fliflet and Read. Use of weakly irregular waveguide theory to calculate eigen frequencies, Q values, and RF field functions for gyrotron oscillators. *Int. Journal of Electronics*, 51, pp-475-484, 1981.

[FRC+82] A.W. Fliflet, M.E. Read, K.R. Chu, and R. Seeley. A self-consistent field theory for gyrotron oscillators: application to a low Q gyromonotrons. *International Journal of Electronics*, 53: pp-505-521, 1982.

[GA82] A.K. Ganguly and S. Ahn. Self-consistent large signal theory of the gyrotron travelling wave amplifier. *International. Journal of Electronics*, 53, pp-641-658, 1982.

[Gap59] A.V. Gaponov. Interaction between electron fluxes and electromagnetic waves in waveguides. Izv. Vyssh. Uchebn. Zaved. Radiofizika, 2, pp-450-462, 1959.

[GC84] A.K. Ganguly and K.R. Chu. Limiting current in Gyrotrons. International *Journal of Infrared and Millimeter Waves*, 5, pp-103-121, 1984.

[GFG+81] Gapanov,A.V., Flaygin,V.A., Gol'denberg, A.L.,Nusinovich, G.S., Tsimring, S.E., Usov,V.G., Vlasov, S.N. Powerful millimeter-wave gyrotrons. *International Journal of Electronics*, 51,277, 1981.

[Gil86] A.S. Gilmour. *Microwave Tubes*, Artech House, Norwood, MA, USA, 1986.

[GLG08] M.Yu. Glyavin, A.G. Luchinin and G.Yu. Golubiatnikov. Generation of 1.5 KW, 1 THz Coherent Radiation from a Gyrotron with a Pulsed Magnetic Field. *Physics Review Letter*, 100, 015101, 2008.

[HG77] J.L. Hirshfield, and V.L. Granatstein. Electron cyclotron maser - An historical survey. *IEEE Transactions. Microwave Theory and Techniques*, 25, pp-522-527, 1977.

[HMH+03] Y. Hirata, Y. Mitsunaka, K. Hayashi, Y. Itoh, K. Sakamoto and T. Imai. The design of a tapered dimple-type mode converter/launcher for high-power gyrotrons. *IEEE Transactions on Plasma Science*, vol. 31, no. 1, pp-142-145, 2003.

[HGA+09] J.P. Hogge, T. P. Goodman, S. Alberti, F. Albajar, K. A. Avramides, P. Benin, S. Bethuys, W. Bin, T. Bonicelli, A. Bruschi, S. Cirant, E. Droz, O. Dumbrajs, D. Fasel, F. Gandini, G. Gantenbein, S. Illy, S. Jawla, J. Jin, S. Kern, P. Lavanchy, C. Lievin, B. Marletaz, P. Marmillod, A. Perez, B. Piosczyk, I. Pagonakis, L. Porte, T. Rzesnickl, U. Siravo, M. Thumm and M. Q. Tran. First experimental results from the European Union 2-MW coaxial cavity ITER gyrotron prototype. *Fusion Science and Technology*, 55, pp-204-212, 2009.

[IKP96] C.T. Iatrou, S. Kern, A.B. Pavelyev. Coaxial cavities with corrugated inner conductor for gyrotrons. *IEEE Transactions on Microwave Theory and Techniques*,Vol. 44, No. 1, 56-64, 1996.

[Ill97] S. Illy. *Untersuchungen von Strahlinstabilitäten in der Kompressionszone von Gyrotron-Oszillatoren mit Hilfe der kinetischen Theorie und zeitabhängiger Particle-in-Cell Simulation.* Forschungszentrum Karlsruhe, Karlsruhe, Germany. Scientific Report 6037, 1997.

[Jel00] John Jelonnek. *Untersuchung des Lastverhaltens von Gyrotrons.* PhD thesis, Technical University Hamburg-Harburg, Dusseldorf, 2000.

[JTP+09] Jianbo Jin, Manfred Thumm, Bernhard Piosczyk, Stefan Kern, Jens Flamm, and Thomasz Rzesnicki. Novel numerical method for the analysis and synthesis of the fields in highly oversized waveguide mode converters. *IEEE Transactions on Microwave Theory and Techniques*, Vol. 57, No. 7, 1661-1668, July 2009.

[JPT+06] J. Jin, B. Piosczyk, M. Thumm, T. Rzesnicki, and S. Zhang. Quasi-optical mode converter/mirror system for a high power coaxial-cavity gyrotron. *IEEE Transactions on Plasma Science*, vol. 34, no. 4, pp-1508-1515, 2006.

[KAB+08] Stefan Kern, Konstantinos A. Avramides, Matthias H. Beringer, Olgierd Dumbrajs, Ying-hui Liu. Gyrotron Mode Competition Calculations: Investigations on the Choice of Numerical Parameters, *33rd International Conference on Infrared, Millimeter and Terahertz Waves (IRMMW-THz 2008)*, Pasadena, CA, USA, 15-19 September 2008, W5D29.1411.

[KAC+10] S. Kern, A. A. Konstantinos, A. R. Choudhury, O. Dumbrajs, G. Gantenbein, S. Illy, A. Samartsev, A. Schlaich, and M. Thumm. Simulation and experimental Investigations on Dynamic After Cavity Interaction (ACI). *International Conference on Infrared, Millimeter and THz Waves*, Rome, Italy, 5-10, 2010.

[KBT04] M.V. Kartikeyan, E. Borie and M. Thumm. *Gyrotrons-High power microwave and millimeter wave technology*, Springer, Berlin,

[KST+08] A. Kasugai, K. Sakamoto, K. Takahashi, K. Kajiwara and N. Kobayashi. Steady-state operation of 170 GHz 1 MW gyrotron for ITER, *Nuclear Fusion*, 48: 054009, 2008.

[KDS+85] K.E. Kreischer, D.G. Danly, J.B. Schutkeker, and R.J. Temkin. The Design of Megawatt Gyrotrons. *IEEE Transactions on Plasma Science*, vol.13, pp-364-373, 1985.

[Ker96] S. Kern. *Numerische Simulation der Gyrotron-Wechselwirkung in koaxialen Resonatoren.* PhD Thesis, University of Karlsruhe, Scientific report FZKA 5837, 1996.

[KSF+09] S. Kern, A. Schlaich, J. Flamm, G. Gantenbein, G. Latsas, T. Rzesnicki, A. Samartsev, M. Thumm and I. Tigelis, Investigations on parasitic oscillations in megawatt gyrotrons, *34th International Conference on Infrared, Millimeter, and Terahertz Waves*, Busan, Korea, 21-25, 2009.

[KT80] K.E Kreischer and, R.J. Temkin. Linear theory of an electron cyclotron maser operating at the fundamental. *International. Journal of Infrared and Millimeter Waves*, 1, pp-195-233, 1980.

[KR81] K. J. Kim and M. E. Read. Design considerations for a megawatt CW gyrotrons. *International Journal of Electronics*, vol. 51, pp-427-432, 1981.

[KKD+97] K. E. Kreischer, T. Kimura, B. G. Danly, and R. J. Temkin. High power operation of a 170 GHz megawatt gyrotron. *Phys. Plasmas*, vol. 4, no. 5, pp-1907-1914, 1997.

[Lit08] G. Litvak, High power gyrotrons. Development and applications, *33rd International Conference on Infrared, Millimeter and Terahertz Waves*, pp. 1-4, 15-19 September 2008.

[Nus04] G. S. Nusinovich, *Introduction to the physics of gyrotrons*, Baltimore and London: The Johns Hopkins University Press, 2004.

[PBD+99] B. Piosczyk, O. Braz, G. Dammertz, C. T. Iatrou, S. Illy, M. Kuntze, G. Michel, and M. Thumm. 165GHz, 1.5MW-coaxial cavity gyrotron with depressed collector. *IEEE Transaction. on Plasma Science.*, vol. 27, no. 2, pp-484-489, 1999.

[PDD+04] B. Piosczyk, G. Dammertz, O. Dumbrajs, O. Drumm, S. Illy, J. Jin, and M. Thumm. A 2-MW 170-GHz Coaxial Cavity Gyrotron. *IEEE Transactions on Plasma Science.*, Vol. 32, No. 3:413-417, 2004.

[PDD+05] B. Piosczyk, G. Dammertz, O. Dumbrajs, S. Illy, J. Jin, W. Leonhardt, G. Michel, O. Prinz, T. Rzensnicki, M. Schmid, M. Thumm and X. Yang. A 2 MW, 170 GHz coaxial Cavity Gyrotron-experimental verification of the design of main components. *Journal of Physics: Conference Series.*, Vol. 25, 24-32, 2005.

[Pio93] B. Piosczyk. Electron guns for gyrotron applications. Edgcombe, C.J. (Hrsg.). *Gyrotron Oscillators - Their Principles and Practice, Chapter 5*, London: Taylor and Francis, pp-123-146, 1993.

[PTV+92] W.H. Press, S.A. Teukolsky, W.T. Vetterling and B.P. Flannery. *Numerical recipes*, Cambridge: Cambridge University Press. 1992.

[RPD+07] T. Rzesnicki, B. Piosczyk, G. Dammertz, G. Gantenbein, M. Thumm and G. Michel. 170GHz, 2 MW coaxial cavity gyrotron - investigation of the parasitic oscillations and efficiency of the RF-output system. *Proceedings of the IEEE International Vacuum Electronics Conference, Kitakyushu, Japan*, pp-45-46, 2007.

[Sch59] J. Schneider. Stimulated emission of radiation by relativistic electrons in a magnetic field. *Physical Review Letters*, 2, pp-504-505, 1959.

[Sch68] F Scheid. *Theory and problems of numerical analysis*, London: McGraw-Hill, 1968.

[SCG+11] Andreas Schlaich, A. Roy Choudhury, Gerd Gantenbein, Stefan Illy, Stefan Kern, Christophe Liévin, Andrey Samartsev, and Manfred Thumm. Examination of Parasitic After-Cavity Oscillations in the W7-X Series Gyrotron SN4R. *International Conference on Infrared, Millimeter and THz Waves*, Houston, TX, USA, 2011.

[Siv65] D.V. Sivukin. *Review of Plasma Physics*. Leotowich, Springer, New York, USA, 1965.

[SN09] O. V. Sinitsyn, and G. S. Nusinovich. Analysis of aftercavity interaction in gyrotrons. *Physics of Plasmas*, 16, 023101, 2009.

[SNA10] O. V. Sinitsyn, G. S. Nusinovich, and T. M. Antonsen. Possibilities for reducing the aftercavity interaction effect in gyrotrons. *Phys.Plasmas 17, 083106 (2010); doi: 10.1063/1.3469577.*

[TAB+03] M. Thumm, A. Arnold, E. Borie, G. Dammertz, O. Drumm, R. Heidinger, M.V. Kartikeyan, K. Koppenburg, A. Meier, B. Piosczyk, D. Wagner and X. Yang. Development of frequency step tunable 1.0 MW gyrotrons in D-Band. *Proceedings of the IEEE International Vacuum Electronics Conference*, Seoul, Korea, 30-31, 2003.

[Thu13] Manfred Thumm. *State-of-the-art of High Power Gyro- Devices and Free Electron Maseres Update 2012*. KIT Scientific Reports 7641, Karlsruhe Institute of Technology, Germany, 2013.

[Twi58] R.Q. Twiss and J.A. Roberts. Electromagnetic radiation from electrons rotating in an ionized medium under the action of a uniform magnetic field. *Australian Journal of Physics*, 11, 424, 1958.

[VZO+69] S.N. Vlasov, G.M. Zhislin, I.M. Orlova, M.I. Petelin, and G.G. Rogacheva. Irregular Waveguides as Open Resonators. *Radiophysics and Quantum Electronics*, Vol. 12, pp-972-978, 1969.

[VZO76] S.N. Vlasov, L.I. Zagryadskaya, I.M. Orlova. Open coaxial resonators for gyrotrons. *Radio Engineering and Electronics Physics.* 21, 96-102, 1976.

[ZM04] V. E. Zapevalov and M.A. Moiseev. Influence of Aftercavity Interaction on Gyrotron Efficiency. *Radiophysics and Quantum Electrodynamics*, 47, pp-520-527, 2004.

Acknowledgments

The present work was accomplished during my Ph.D. programme, at the Institute for Pulsed Power and Microwave Technology (IHM) at Karlsruhe Institute of Technology (KIT) in the frame of DAAD Fellowship. During this period a number of people contributed towards its making successful and I have the deepest pleasure hereby in acknowledging their efforts.

First, and foremost, I would like to extend my deepest gratitude and thanks to Prof. Dr. rer. nat. Dr. h.c. Manfred Thumm, for his illuminative guidance, academic suggestions, continuous efforts and encouragement throughout this period and for being my doctoral adviser. Without his support it could not have seen the light of success. He was always helpful in solving technical as well as personal problems. I feel privileged to have been his student and to have had the chance to learn from his expertise in the field of gyrotron, as well as to interact with his personality. His confidence in me and the freedom he gave me helped me to accomplish this work.

I am grateful to Prof. Dr.–Ing. Dr. h.c. Klaus Schï¿½nemann from TU Hamburg–Harburg for his great interest in my work and being my second reviewer. I also thank Prof. Dr.–Ing.John Jelonnek for his excellent guidance and thorough discussions. I wish to express my sincere gratitude to Dr. Stefan Kern for the immeasurable discussions on solving my Ph.D. problem and also for his valuable discussions and feedback. I gratefully acknowledge the Head of the Division, Dr. Gred Gantenbein and Dr. Bernhard Piosczyk for his friendship and technical advices. I am delighted to acknowledge Dr. Danilo D' Andrea for providing me the necessary support and technical advice during the entire period of my stay at IHM/KIT, without his support, my work would not have been possible. Furthermore, I would like to thank Prof. Edith Borie, Dr. Stefan Illy and Dr. Konstantinos A. Avramidis for helping me out from time to time. I want to express my deepest appreciation to Dipl.–Ing.Andreas Schlaich for his co–operation and inspirational technical discussions. I would also like to thank all my colleagues at IHM/KIT for the warm hospitality and providing me a very co-operative ambience, in which, completing my work was a privilege and pleasure.

I am thankful to Dr. Olgerts Dumbrajs, Institute of Solid State Physics, University of Latvia, for providing me the necessary technical advice and suggestions.

Author expresses his indebtedness to Dr. Chandra Shekhar (Director, CSIR–CEERI Pilani) and Dr. S.N. Joshi (Emeritus Scientist, CSIR–CEERI Pilani) for providing all necessary support and guidance throughout his research work. Without their help it would be impossible to enhance his academic carrier in terms of getting Ph.D. degree from KIT Germany.

Author also expresses his sincere thanks to Dr.Vishnu Srivastava (Head of the Division, MWT), Dr. R.S. Raju, Dr. A.K. Sinha, Dr. L.M. Joshi, Er. R.K.Gupta, Dr.R.K.Sharma and all other members of the MWT Division of the CSIR–CEERI Pilani for their encouragement and continuous support.

My best thanks goes to my friends Uttam Goswami, CSIR-CEERI Pilani and Parth C. Kalaria, Visiting Scientist, ITER-India for their great co–operation and support from time to time.

I would like to extend my deepest gratitude to Prof. Dr. B.N. Basu for his excellent support during the whole tenure of my Ph.D. work. I would also like to thank him for his encouragement and guidance.

I would like to extend my deepest gratitude to Prof. M.V. Kartikeyan, IIT Roorkee for his friendship and his constant encouragement to carry out this work. I would like to thank him for providing me valuable technical discussions when he was in IHM/KIT.

My very special thank goes to family, especially to my mother, who was always there for me. Without her, I would not be where I am today. I also thank my father for his tireless support and encouragement. I wish to extend my sincere thanks to my in–laws for providing a loving environment for me whenever it was necessary, particularly to my father–in–law for his support from time to time. Furthermore I also thank my elder brother Avijit Roy Choudhury, Scientist ISRO for encouraging me to do my Ph.D. more successfully.

Above all, no words would be sufficient to express my gratitude to my wife Dr. Tanami Roy for her love, continuous support and for her sacrifice by giving me the time and space I needed to accomplish this work. Her constant help, co–operation and inspiration made it possible to complete my work. I am very lucky to have her by my side.

I am forever indebted to all of them for their unconditional love, understanding, endless patience and encouragement when it was most required.

Finally, I would like to thanks to my sponsor Deutscher Akademischer Austauschdienst–DAAD (German Academic Exchange Service) for providing me the necessary financial assistance.

Karlsruhe, May 2013
Amitavo Roy Choudhury